Samuel Slater and the Origins of the American Textile Industry, 1790–1860

BARBARA M. TUCKER

CORNELL UNIVERSITY PRESS

ITHACA AND LONDON

First published 1984 by Cornell University Press.
Published in the United Kingdom by
Cornell University Press Ltd., London.

International Standard Book Number 0–8014–1594–2
Library of Congress Catalog Card Number 84–45145
Printed in the United States of America
*Librarians: Library of Congress cataloging information
appears on the last page of the book.*

*The paper in this book is acid-free and meets the guidelines
for permanence and durability of the Committee on Production
Guidelines for Book Longevity of the Council on Library Resources.*

For my parents

Contents

[7]

Illustrations

[9]

Tables

Tables

Preface

Until recently historians of the Industrial Revolution have been slow to extend their investigations into the lives of American workers beyond the scope of the trade-union movement. Although they now have begun to take notice of the culture of artisans—shoemakers, machinists, and hatmakers, for example—strike activity, class formation, and the trade-union movement continue to command a major share of attention.[1]

Interest in the artisan has left the great majority of workers relatively unexplored. Yet skilled workers, who represented at most from 12 to 14 percent of the entire industrial labor force in the nineteenth century, do not accurately reflect the aspirations and concerns of working people as a whole. The experience of a machinist or shoemaker was not the same as that of a weaver, straw-bonnet maker, or machine operator.[2] By concentrating on the skilled

1. David Montgomery, "To Study the People: The American Working Class," *Labor History* 21 (Fall 1980): 489; Alan Dawley, *Class and Community: The Industrial Revolution in Lynn* (Cambridge, Mass., 1976); Paul G. Faler, "Workingmen, Mechanics, and Social Change: Lynn, Massachusetts, 1800–1860," Ph.D. dissertation, University of Wisconsin, 1971, and *Mechanics and Manufacturers in the Early Industrial Revolution: Lynn, Massachusetts, 1780–1860* (Albany, 1981); David Harlan Bensman, "Artisan Culture, Business Union," Ph.D. dissertation, Columbia University, 1977; Shelton Stromquist, "Enginemen and Shopmen: Technological Change and the Organization of Work in an Era of Railroad Expansion," paper presented at the 74th annual meeting of the Organization of American Historians, Detroit, April 2, 1981; Judith McGaw, "The Sources and Impact of Mechanization: The Berkshire County, Massachusetts, Paper Industry, 1801–1885 as a Case Study," Ph.D. dissertation, New York University, 1977.

2. Andrew Dawson, "The Paradox of Dynamic Technological Change and the Labor Aristocracy in the United States, 1880–1914," *Labor History* 20 (Summer

craftsman's response to industrialization, by making him the chief spokesperson for the entire laboring population, and by focusing considerable attention on conflict with management, historians have ignored many of the nuances and much of the diversity of the lives led by working-class men and women.

Although recently an increasing number of labor historians have begun to examine the response of unskilled laborers to industrialization, these works, too, are concerned primarily with conflict, class formation, and labor organization.[3] In a study of antebellum New England factory life, a recent scholar argued that "opposition was intrinsic to the nature of industrialization," that labor and management were persistently at odds, and that their struggles "produced what can be properly viewed as a class relationship between factory employees and employers."[4] Class conflict and class consciousness are not a given even if we accept the fact that classes existed. In any industrial situation the potential for overt class conflict is undeniably present, but so is the potential for cooperation among the classes. An exclusive focus on conflict—or undercurrents of conflict, or "repressed struggle"—between labor and management not only distorts the historical record, it also is unfaithful to the actual experience of workers.

It is a gauge of the advance made by labor historians in recent years that one must be a revisionist to examine the role of stability and cooperation between labor and management and their integration in the new industrial order. By "stability" I do not imply an absence of tension. In factories and in industrial communities, change was a feature of social and economic life, and strains were

---

1979):325–351. There are exceptions: see Tamara K. Hareven, *Family Time and Industrial Time* (New York, 1982); Virginia Yans-McLaughlin, *Family and Community: Italian Immigrants in Buffalo, 1880–1930* (Ithaca, 1977).

3. Daniel J. Walkowitz, *Worker City, Company Town: Iron and Cotton Worker Protest in Troy and Cohoes, New York, 1855–1884* (Urbana, 1978); Bruce Laurie, *Working People of Philadelphia, 1800–1850* (Philadelphia, 1980); Thomas Dublin, *Women at Work: Transformation of Work and Community in Lowell, Massachusetts, 1826–1860* (New York, 1979); Jonathan Prude, *The Coming of Industrial Order: Town and Factory Life in Rural Massachusetts, 1810–1860* (New York, 1983); Gary B. Kulik, "Pawtucket Village and the Strike of 1824: The Origins of Class Conflict in Rhode Island," *Radical History Review* 17 (Spring 1978):5–37; see also Gary B. Kulik, "The Beginnings of the Industrial Revolution in America: Pawtucket, Rhode Island, 1672–1829," Ph.D. dissertation, Brown University, 1980.

4. Prude, *Coming of Industrial Order*, p. xii.

evident between labor and management. Yet the disagreements that did occur did not alter or indeed threaten appreciably the new order, nor did they necessarily produce class confrontation or indicate the emergence of class consciousness.

My purpose is to explore the persistence of traditional culture amid the rapid technological and economic changes of the early Industrial Revolution in New England, focusing on working-class culture in the context of the general social and economic trends of the late eighteenth and early nineteenth centuries. To study the lot of the worker without at the same time gaining an understanding of the manufacturer, the industry, and the overall economy would be to draw a false impression. Labor and management alike are influenced by the prevailing values and institutions of society and by objective circumstances in industry and in the economy. The relationship between labor and management in the early Industrial Revolution was not a simple one, and not inevitably an antagonistic one.

This book is largely a case study of Samuel Slater's factory system between 1790 and 1860. In Part I, I examine the early British system of textile manufacture and Slater's attempt to introduce aspects of that system to America. At his Pawtucket mill, Slater experimented with a variety of ownership, management, and labor schemes, some of them borrowed from his British past, others suggested either by his partners, William Almy and Smith (and, later, Obadiah) Brown, or by the householders in his work force. Part II traces the growth of the textile industry and Samuel Slater's contribution to the process of industrial development. Slater eventually became one of the most important factory masters of the nineteenth century; he owned factories and firms throughout New England, particularly in southern Massachusetts, Rhode Island, and Connecticut. Webster, Massachusetts, developed into a showcase for his factory system, and it was there that the force of tradition was most clearly apparent: the allocation of work, the discipline of hands, the payment of wages, the "accommodation," the social institutions, even the design of the town conformed to customary patterns.

Many of the factory towns established during these years appeared to be peaceful, ordered, traditional communities to those who lived and worked in them. The integration of customary preindustrial institutions, authority patterns, and beliefs within the new industrial order allowed a spirit of cooperation to prevail between employer and factory householder and helped to shape an economic system that served their respective needs.

Although cooperation characterized the relationship between Slater and his workers for a generation or so, the potential for conflict is clearly inherent in many of the situations I discuss. Flight, the reaction of most Slater employees to the slow dissolution of traditional prerogatives, is arguably the most resolute form of protest. And tensions long embedded in customary relationships between men and women, parents and children, the churched and the unchurched surfaced in this new setting. In the long run the benign, paternalistic structure of New England society was unable effectively to assimilate massive economic change. Increased competition, the growth of the market economy, the ready availability of new and inexpensive sources of labor, and the ascension of Slater's sons to power in the family firm caused this fragile social structure to disintegrate. By the 1830s a new relationship between labor and management had begun to emerge. These themes are taken up in Part III.

In the years required to research and write this book, I have had the assistance of many people and organizations, and it is with sincere appreciation that I acknowledge their interest, encouragement, enthusiasm, and support for this project.

A fellowship from the Leverhulme Commonwealth/American Foundation allowed a year of research in England at the University of Sheffield; research grants from both Sydney University and Rutgers University supported travel and other expenses. Portions of this book have appeared in earlier versions as "The Merchant, the Manufacturer, and the Factory Manager: The Case of Samuel Slater," *Business History Review* 50 (Autumn 1981): 297–313; "'Our Good Methodists': The Church, the Factory, and the Working Class in Ante-Bellum Webster, Massachusetts," *Maryland Historian* 8 (Fall 1977): 26–37; and "The Family and Industrial Discipline in Ante-Bellum New England," *Labor History* 21 (Winter 1979–80): 55–74. I thank the editors of the respective journals for permission to use this material here.

Preface

The residents of Webster shared my enthusiasm for their town's past. Especially helpful were Hugh Ward Miller, then pastor of the United Church of Christ, Federation, who directed my attention to the rich collection of Methodist and Congregational documents owned by his church; Mrs. S. Galvin and town officials answered numerous questions on their community and its past. Richard Candee helped me to understand and appreciate mill architecture and community design, and colleagues at Rutgers University, including Paul G. E. Clemens, Suzanne Lebsock, Richard L. McCormick, and James Reed, read all or parts of the manuscript and made important suggestions on both content and analysis. I am especially grateful to Mary Hartman, who took time from her own study of gender to clarify my ideas on patriarchy. Stephen Salsbury of Sydney University excited my interest in business history and provided valuable comments on Slater's business practices. Daniel H. Calhoun and Lloyd Gardner supplied the intangible ingredients necessary in any undertaking of this magnitude—encouragement and support. Those fortunate enough to know these men acknowledge their intellectual abilities and applaud their integrity and humanity. Kenneth H. Tucker, Jr., deserves special mention. He has come to know and respect the workers of Webster; his insights illuminate many of the ideas presented here, and he contributed much to whatever grace this book possesses.

David Brody has read innumerable versions of this manuscript and has proved to be a critical, demanding, yet enthusiastic and sympathetic guide. He first suggested that the Slater papers would provide an exciting and valuable area for research but cautioned that it would defy easy, quick, or conventional analysis. He was right.

The memory of Titiana Kashperevitz Lakedon, my grandmother, infuses this work. For forty-seven years she worked in the textile mills of Rhode Island and Massachusetts; she knew what was important and taught her children and grandchildren.

Barbara M. Tucker

New Brunswick, New Jersey

Samuel Slater and the Origins
of the American Textile Industry,
1790–1860

Introduction

In the United States industrialization came initially to the manufacture of textiles. Samuel Slater established the earliest successful American spinning mill at Pawtucket, Rhode Island, in 1790. In establishing it and his other mills Slater adapted knowledge he had gained from factory masters in his native Derbyshire, England, to meet the special needs of American unskilled laborers and local merchants. Industrialization, under Slater's guidance, did not bring the immediate destruction of the old New England values, beliefs, institutions, and authority patterns. In colonial New England the groups that linked individuals in society and shaped their identity were the family, the church, and the community. These institutions and the patterns of authority and hierarchy manifest in them became the foundation on which Samuel Slater erected his factory system. The patriarchal family structure; the customary division of labor on the basis of age, gender, and marital status; the church; and the New England town design were integral to many of the early factory colonies. Such respect for regional culture served to ease the transition, for all elements of the community, from an agricultural, rural economy to one based on the factory system and wage labor. The early factory system established by Samuel Slater served initially the needs and requirements of both employer and employee.

Historians generally agree that seventeenth-century society was patriarchal. Notions of subordination, obedience, and reciprocal duties and responsibilities defined the relationship between husband and wife, parents and children, and masters and servants. The basis for male authority within the family consisted of two intertwined components: economic power and what might be termed social or moral power. Men had a duty to provide for their families. Ownership of property often guaranteed male authority. Recognizing that power lay in domination of this resource, men only reluctantly, if at all before their deaths, relinquished control over their property.[1] But beyond the provision of food and shelter, a man had other, equally important responsibilites, especially concerning the moral and educational upbringing of children and servants. Parents had a duty to see that children would be able to earn a living.

1. Perhaps the best-known advocate of patriarchy was Sir Robert Filmer, seventeenth-century philosopher and defender of the "divine right of kingship." In his influential work *Patriarcha*, published in 1690, Filmer used the fatherly image as the basis of political obligation and as an explanation for the origins of political power. Using the Bible as the ultimate authority on the nature of society, he argued that the family was the natural form of social organization. Within that unit paternal power was absolute; restrictions against paternal authority were indefensible. Fathers ruled as successors to Adam and shared with him the power God had given. "God had specifically granted all social power to His direct ancestor, Adam, and He had created the social order in such a way that the continuation of the hierarchy founded by Adam was indispensable." Filmer transferred this argument to politics to justify absolute monarchy.

> If we compare the natural duties of a Father with those of a King we find them to be all one, without any difference at all but only in the latitude of extent of them. As the Father over one family so the King, as Father over many families, extends his care to preserve, feed, clothe, instruct and defend the whole commonwealth. His wars, his peace, his courts of justice, and all his acts of sovereignty, tend to preserve and distribute to every subordinate and inferior Father, to their children, their rights and privileges, so that all the duties of a King are summed up in an universal father care of his people.

But the king did not require consent from the families he governed. In society as "in every family the government of one alone is most natural." See Peter Laslett, ed., *Patriarcha and Other Political Works of Sir Robert Filmer* (Oxford, 1949), pp. 18, 63, 84.

Filmer's work codified widely held assumptions about family, society, the economy, and politics. While John Locke later argued that the state and the family rested on different bases, little was done to challenge the social and economic implications of patriarchy. The patriarchal family characterized seventeenth-century society. The father dominated the family and expected obedience and deference from wife and children. But as Filmer's statement implies, the father had the specific duty to provide his family with food, shelter, and clothing and to protect their interests, to prepare his children for a calling and to instruct them in religious principles and ethics. Beyond the family, patriarchy influenced wider economic values, institutions, and practices. Private property, primogeniture, and the subordination of the economic interests of women and children to those of men are based in part on this concept. See Laslett, *Patriarcha*, p. 23; see also Philip J. Greven, Jr., *Four Generations: Population, Land, and Family in Colonial Andover, Massachusetts* (Ithaca, 1970), pp. 72–99.

A man did not fulfill his obligations until he saw his children "well dispos'd of, well settled in the World," and until he taught them to be responsible, moral citizens.[2] Lessons in piety, obedience, reverence, and deference, as well as lessons in reading and writing, formed the basis of education and discipline.[3] In the seventeenth century fulfillment of both economic and moral obligations brought self-respect, ensured men the esteem of their families and their communities, and gained them status in society. In this structure, women, children, and servants remained clearly subordinate. Women, for example, were cautioned by the clergy to "guid the house &c. not guid the Husband."[4] And they should submit to their husbands' instructions and commands: "He stood before her in the place of God: he exercised the authority of God over her, and he furnished her with the fruits of the earth that God had provided."[5]

Evidence from the early eighteenth century indicates that while the relationship between husband and wife, parent and child, or householder and servant was more elastic than it had been a century earlier, family members still lived and worked within the limits set by the male head of the household. Laurel Thatcher Ulrich, who explored the relationship between husband and wife in eighteenth-century New Hampshire, concluded that in the economic sphere "almost any task was suitable for a woman as long as it furthered the good of her family and was acceptable to her husband. The approach was both fluid and fixed. It allowed for varied behavior without really challenging the patriarchal order of society."[6] Children and servants, too, lived within boundaries prescribed by the male householder.[7]

While by extension patriarchy infused all aspects of New En-

2. Benjamin Wadsworth, *The Well-Ordered Family* (Boston, 1712), quoted in Edmund S. Morgan, *The Puritan Family: Religion and Domestic Relations in Seventeenth-Century New England*, (New York, 1966 [1944]), p. 79.

3. Peter Sterns, *Be a Man! Males in Modern Society* (New York, 1979), pp. 25–49.

4. Boston Sermons, September 30, 1672, quoted in *Puritan Family*, p. 43.

5. Ibid., p. 45.

6. Laurel Thatcher Ulrich, *Good Wives: Image and Reality in the Lives of Women in Northern New England, 1650–1750* (New York, 1982), pp. 37–38. See also Nancy Folbre, "Patriarchy in Colonial New England," *Review of Radical Political Economics* 12 (Summer 1980):4–11; Alice Kessler-Harris, *Out to Work: A History of Wage-Earning Women in the United States* (New York, 1982), pp. 3–19.

7. Peter G. Slater, *Children in the New England Mind in Death and in Life* (Hamden, Conn., 1977), pp. 114, 127.

gland life in the seventeenth and early eighteenth centuries, merchants, artisans, and unskilled laborers developed different conceptions of patriarchy which corresponded to their economic and social positions. Merchants and traders of the major northern port cities were strongly committed to a form of patriarchy that ensured that the family was the basic business unit. Kin organized partnerships, pooled capital and resources, and participated in the active management of their firms. Ownership and management went hand in hand. Few merchants entrusted appreciable authority to outsiders, believing that waste, fraud, mismanagement, and theft would be the result. Family connections, rather than proved ability, determined the choice of managers, supercargoes, and foreign agents.[8]

Powerful international family partnerships emerged in the colonial era. One such family was the Browns of Providence, who financed the Pawtucket spinning mill. From headquarters in Rhode Island, Obadiah Brown and his four nephews, Nicholas, Joseph, John, and Moses, traded within a worldwide network of kin. In Liverpool the Brown family was represented by Brown, Shipley and Company; in New York by Brown Brothers and Company; in Philadelphia by Brown and Bowen; and in Baltimore by Alexander Brown and Sons. Among such businessmen, family ties determined trading arrangements. If sons, nephews, or other kin wanted to share in the business, they were bound to follow the dictates of the paternal head, who determined the type of education they received, their religious training, often the age at which they would marry and enter the business, and the jobs they would perform. Within the family unit merchants exercised both moral and financial authority.[9]

For craft workers in the seventeenth and eighteenth centuries, knowledge of a skill ensured authority over kin. The home and the workshop were inseparable. Among master shoemakers, for example, the family was an extended household unit composed of wife, children, journeymen, and apprentices, all of whom lived and

8. Alfred D. Chandler, Jr., *The Visible Hand: The Managerial Revolution in American Business* (Cambridge, Mass., 1977), pp. 36–37; Bernard Farber, *Guardians of Virtue: Salem Families in 1800* (New York, 1972), pp. 75–78.

9. James B. Hedges, *The Browns of Providence Plantations: Colonial Years* (Cambridge, Mass., 1952), pp. 13–14, 90–91; Chandler, *Visible Hand*, pp. 36–37.

worked under the direct supervision of the master craftsman and were subject to the "master's paternal and property authority."[10] In this unit the master exercised his authority at once as head of the household and as head of a production team.

While among merchants and skilled workers economic and moral authority merged, power within the family did not necessarily require extensive economic supports. Among unskilled people, the family also was decidedly patriarchal. By the mid-eighteenth century, the economic underpinnings of vast numbers of people had been eroded. Roughly 10 percent, and in some areas even more, of the population owned no real estate or movable assets whatsoever, and few possessed discernible skills. These landless, skill-less men and their families drifted from place to place in search of work and the chance of permanent settlement.[11] Despite the decided blow to their position as primary breadwinner in the family, such men continued to regard themselves as the natural head of the household, and they tried to retain their authority over and responsibility for the education, discipline, training, and protection of wife and children.

Some scholars have suggested that patriarchy was on the wane by the time of the Revolutionary War. A broader understanding of the nature of parental authority was emerging. Parental power and responsibility, filial rights and prerogatives, came under review and were subject to debate. The dominant themes in scores of books, pamphlets, and other works published in Britain and America stressed affection, equality, and a growing sense of autonomy and independence between the generations. The practical effect of these ideas on family government and parental discipline across the economic spectrum is difficult to gauge.[12] An argument can be made

10. Dawley, *Class and Community*, p. 18; see also Faler, "Workingmen, Mechanics, and Social Change," pp. 48–52.

11. Greven, *Four Generations*, pp. 222–258; Douglas Lamar Jones, "Geographic Mobility and Society in Eighteenth-Century Essex County, Massachusetts," Ph.D. dissertation, Brandeis University, 1975, pp. 161–206; Jackson T. Main, *The Social Structure of Revolutionary America* (Princeton, 1965), pp. 22–23; Charles Grant, *Democracy in the Connecticut Frontier Town of Kent* (New York, 1961), p. 97; Kenneth Lockridge, "Land, Population, and the Evolution of New England Society, 1630–1790," *Past and Present* 39 (April 1968):62–80.

12. Jay Fliegelman, *Prodigals and Pilgrims: The American Revolution against Patriarchal Authority, 1750–1800* (New York, 1982), pp. 1–6. Fliegelman admits that it is

that among the upper levels of society, "fond affection rather than conscientious discipline shaped the relationship between the generations."[13] While patriarchy may have been on the decline among some segments of society, its hold over the lower classes survived and was abetted by the new mill system.[14]

The Slater factory system appealed to many unskilled male householders because it allowed them to reinforce their moral and economic authority within the family. Rather than challenge customary prerogatives, the new industrial order protected and even bolstered patriarchy among the lower classes. Negotiations between Slater and householders established the family as the primary source of labor in the new factory system and protected the householders position as provider, guide, and teacher of wife and children. The organization of labor both inside and outside the factory reflected this arrangement. In Slater's communities a traditional division of labor emerged in which men were given jobs as casual workers, farm hands, teamsters, and herders; children provided the labor force for the new mills. Parents did not compete with children for work or wages. In the factory itself the process of work remained in the hands of kinship units, and parents exercised considerable influence over the allocation of jobs, the discipline of hands, and the method of wage payment.

difficult to "demonstrate an immediate or direct causal relationship between a set of ideas and a sequence of political or social events. Such overly insistent arguments are invariably unsatisfying; for the relationship between idea and event is intractably complex" (p. 6). Others have stressed the changes and strain placed on patriarchy. Greven, for example, concluded that by the mid-eighteenth century, "if patriarchalism was not yet gone, it had been made less viable by the changing circumstances. The earlier economic basis which had sustained the attempts by fathers to establish and to maintain their control and influence over the lives of their sons no longer was to be found among the majority of families living in Andover. . . . For the majority, another conception of their roles and their authority as fathers had become necessary" (*Four Generations*, p. 273). While this situation might have obtained for fathers and their adult sons, could a decline in patriarchalism be seen in the primary relations between husband and wife, parents and young children, householder and servant? For other views, see Nancy Cott, "Eighteenth-Century Family and Social Life Revealed in Massachusetts Divorce Records," *Journal of Social History* 10 (1976):131–160; Daniel S. Smith, "Paternal Power and Marriage Patterns: An Analysis of Historical Trends in Higham, Massachusetts," *Journal of Marriage and the Family* 35 (1973).

13. Philip J. Greven, Jr., *The Protestant Temperament: Patterns of Child-Rearing, Religious Experience, and the Self in Early America* (New York, 1977), p. 265.

14. Ibid., p. 32. See also Tamara K. Hareven, *Family and Kin in Urban Communities, 1700–1930* (New York, 1977), pp. 3–4.

[26]

Religion, another feature of New England colonial life, also flourished under the new factory system. The religious history of the area is well documented and does not need to be repeated here. Throughout the course of the colonial era, the region was the scene of spiritual activity and a constant renewal of piety. Puritan enthusiasm dominated the seventeenth century, and religious spirit was kept alive by the Great Awakening and subsequently by scattered local revivals, especially in 1763 and 1764. Toward the close of the eighteenth century, religious intensity surfaced again in many New England towns and churches. Known as the Second Great Awakening, this Christian renewal was sustained for a generation and more by religious revivals that swept the region.[15] Commitment to Christian values was evident not only in established agricultural communities but also in the new industrial villages built at the turn of the century. Not surprisingly, mill towns became centers for the new revivals. Eric Hobsbawm's observation on religion and the lower classes in England could apply equally well to the American scene: ". . . it would be incredible if the forms and fashions of traditional religion, which had enclosed the lives of the common people from time immemorial, were to have suddenly and completely dropped away."[16] Religion remained a potent force in the lives of New Englanders.

Congregationalism was the dominant religion in the area, but toward the end of the eighteenth century its position was challenged by a reorganized Protestant Episcopal church and by the members of two proselytizing evangelical faiths—the Baptists and the Methodists. In Slater's villages two sects took root. In Slatersville the Congregationalists predominated, and in Webster the Methodists constituted the largest denomination. Both sects emphasized order and authority, and both tried to discipline and monitor the behavior of parishioners. Surveillance extended beyond church walls into the home, the workplace, and the leisure activities of communicants.[17] For many workers the church pro-

---

15. Sydney E. Ahlstrom, *A Religious History of the American People* (New Haven, 1972), pp. 280–94, 403–454.
16. E. J. Hobsbawm, *Primitive Rebels: Studies in Archaic Forms of Social Movements in the Nineteenth and Twentieth Centuries* (New York, 1959), p. 127.
17. There were distinctions between the two groups. They held different positions on the education of the clergy, the recruitment and admission of new mem-

vided a sense of security in the new environment. Religious doctrine and discipline served as a prescription for emotional and economic survival.

The links with New England's colonial past were made more secure by Slater's adoption of the traditional rural village as a blueprint for the design of his factory colonies. In the seventeenth century, close-knit villages patterned on the open-field system of town design and land use predominated. Society revolved around the group or community. Although by the eighteenth century the distinction between community and countryside had become blurred as many residents consolidated their landholdings and increasingly built homes on outlying property, still the village remained the center of social, political, and religious life. In these townships a rough system of democracy based on the town meeting emerged. Leadership was not distributed widely throughout the community, but was vested in the most eminent members of the town, men whose worth and status were recognized by most residents.[18] In appearance many of the factory colonies established by Samuel Slater conformed broadly to these early New England villages. The very organization and physical pattern of the factory towns served to calm the anxiety of laborers who moved from one town to another and from one line of work to another. Towering factories and barrackslike boardinghouses, the symbols of Lowell, were slow to appear in Slater's communities.

In colonial society the identity of an individual depended in part

---

bers, and the independence allowed each church. By the eighteenth century, for example, Congregational ministers formed a significantly well-educated professional group. Gone was the notion that the minister was the leader of a fellowship of Christian believers; pastors tried to argue that they were set apart from their congregations, "that they themselves constituted a peculiarly holy group of men— as if in a society gone wrong they alone embodied the religious life of the community" (J. William T. Youngs, Jr., "Congregational Clericalism: New England Ordinations before the Great Awakening," *William and Mary Quarterly* 31 [July 1974]:487). Methodism seems to have been more democratic. Methodists drew their clergy from among the common people. These men operated within a structured, somewhat autocratic system of districts, circuits, and preaching stations. Most of the ministers were lay preachers; only an elite minority was ordained. See Ahlstrom, *Religious History of the American People*, pp. 280–294, 403–454.

18. John W. Reps, *The Making of Urban America: A History of City Planning in the United States* (Princeton, 1965), p. 125; Edward M. Cook, Jr., *The Fathers of the Town: Leadership and Community Structure in Eighteenth-Century New England* (Baltimore, 1976), chaps. 1 and 2.

on his or her position in the family, the church, and the community. It shaped the way he or she perceived the world, related to other people, and responded to change. Within the context of the new industrial order, customary alliances and values continued to shape lives.

Custom also supported the needs of early factory masters. Traditional values and institutions not only attracted workers to factory colonies; the pattern of authority and hierarchy evident in them became the basis for labor–management relations. In Slater's communities the exercise of power came to rest on preindustrial patterns of authority. Its dimensions could be sketched as follows: Authority relations always involved domination and subordination. Superiority in the family lay in the domination of husband over wife, parents over children, and master over servant; in the church, minister over congregation; and in the community, officials over residents. These individuals or units had specific duties and responsibilities to perform, and failure to execute obligations could result in a challenge to their position and perhaps in an abrogation of reciprocal responsibilities. The duties of various individuals and units, however, were loosely drawn and subject to repeated challenge, testing, and renegotiation. Whether in family, church, or community, the basis of obedience to authority implied more than fear and coercion; implicit was a moral obligation to obey. In all of these situations, power was exercised by face-to-face contact and often in small group situations.[19] Because he based his system of labor-management relations on these precepts, laborers easily accepted his control; custom conferred a degree of legitimacy on his power and position. Once the system was set in place and accepted by labor, it became part of the natural order. Yet in the end labor's adherence to customary institutions and its reluctance to alter traditional patterns of behavior could be used by employers to manipulate and control workers.

By retaining traditional patterns of authority, values, and institutions, Slater also effectively discouraged the formation of new alliances and the development of new identities. Conflict in this society centered primarily around tension embedded in the colo-

19. Barrington Moore, Jr., *Injustice: The Social Bases of Obedience and Revolt* (White Plains, 1978), pp. 16–24.

[29]

nial order: issues associated with gender, family, and religion divided people and overrode class issues. Together labor and management forged a factory system that met their respective requirements. This arrangement was the product of specific social, cultural, and economic forces. It lasted for a generation, and then the economic changes of the 1830s swept away a culture that did not go along with them.

PART I

THE ERA OF EXPERIMENTATION

[1]

British Industrialization: A
Model for American Manufacturers?

American industrialization was part of a process that by 1860 had transformed much of Western society. The industrial development of Belgium, France, Germany, and the United States was heavily indebted to British factory masters. Britain was the first nation to industrialize, and its technology, managerial procedures, labor practices, industrial architecture, and even the design of its factory communities were imitated by scores of industrialists elsewhere. By and large, historians of early American industrialization have argued that the Slater system in particular was an offshoot of the British experience, and therefore Slater's factories often have been dismissed as being essentially foreign. They were not, it is asserted, genuine outgrowths of indigenous social and economic conditions. Instead, the Lowell system, which was used extensively throughout northern New England, has been taken to represent the legitimate precursor of the American business system.[1] But the Slater system was not a root-and-branch transplant from the Old World. To understand in what way it was unique, we must first examine its British antecedents.

For centuries the commercial production of textiles in Great

1. Caroline F. Ware, *Early New England Cotton Manufacture: A Study in Industrial Beginnings* (Boston, 1931), pp. 60, 74; See also Sidney Pollard, "Industrialization and the European Economy," *Economic History Review* 26, ser. 2 (November 1973): 636–648; Reinhard Bendix, *Work and Authority in Industry* (New York, 1956), chap. 2; David S. Landes, *The Unbound Prometheus: Technological Change and Industrial Development in Western Europe from 1750 to the Present* (Cambridge, Eng., 1969), chaps. 2 and 3.

Britain had been handled by merchant manufacturers who operated vast putting-out networks. They bought raw materials and distributed them to spinners who worked, often only part time, spinning wool, flax, or cotton at home. The manufacturers examined the yarn that was produced and then put it out to another set of workers, who wove it into cloth. In this way the manufacturers were able to dictate the styles to be produced, the wages to be paid, and the prices to be charged for the finished goods. Yet despite their best managerial efforts, operations seldom ran smoothly. Most merchant manufacturers did not know when their outworkers would return the yarn and cloth, or indeed if the workers would return it at all. Theft posed a serious problem with outworkers, as did embezzlement and the uneven workmanship. Some spinners could not spin the fine counts of yarn that were occasionally required, and weavers often returned cloth that was thin, uneven, dirty, or short; both kinds of problems occasioned waste and delay.[2]

Further deficiencies in the system appeared in the 1760s, when the demand for cloth increased. Because they could not easily increase the amount of yarn that was spun, merchant manufacturers found it difficult to take advantage of the expanding market. In an effort to remedy this problem, groups and individuals encouraged the development of mechanical yarn-spinning devices. The calico printers Robert Peel, William Yates, and Jonathan Howarth employed James Hargreaves to fashion equipment for them. He constructed a carding cylinder in 1760 and, four years later, a spinning machine known subsequently as the jenny.[3]

2. Sidney Pollard, *Genesis of Modern Management: A Study of the Industrial Revolution in Great Britain* (London, 1965), pp. 31–34; Alfred Wadsworth and Julia De L. Mann, *Cotton Trade and Industrial Lancashire, 1600–1780* (Manchester, Eng., 1931), pp. 274, 282, 405, 472; Sidney Pollard, "Fixed Capital in the Industrial Revolution in Britain," *Journal of Economic History* 24 (September 1964): 307.

3. In 1761 the Society of Arts advertised for "the best invention of a machine that will spin six threads of wool, flax, hemp or cotton, at one time, and that will require but one person to work and attend it." A succession of inventors submitted their hardware in hopes of obtaining the £50 premium offered, but not one solved the problems satisfactorily. More successful were the efforts of merchant manufacturers. See Society of Arts, "Minutes of Transactions 1762–1763," quoted in William A. Hunter, "James Hargreaves and the Invention of the Spinning Jenny," *Newcomen Society: Transactions* 28 (1951–1952 and 1952–1953):144. After Hargreaves proved the success of his machines, he constructed about twenty of them for his employers. This simple, inexpensive, hand-operated machine was adopted by scores of manufacturers. In seven Midland counties approximately one hundred mills were estab-

Following closely on the invention of the jenny came Richard Arkwright's water-frame and carding machine. First built in Preston in 1768, the Arkwright spinning machine possessed several advantages over the simple jenny: a single worker could spin several threads at once, producing yarn that was strong, uniform in thickness, and perfect for warps. Furthermore, it was powered by horses or by water rather than by hand. Arkwright first installed his horse-driven machinery in a reconverted house in Nottingham, but it did not remain there long. With the financial assistance of two local merchant manufacturers, Arkwright transferred operations to the small rural village of Cromford, where he developed all the elements of his factory system.[4]

To house and operate his equipment, he constructed a five-story stone structure and installed a water wheel for power. By 1777 he employed two hundred people, mostly women and children drawn from families whose primary livelihood derived from the local lead-mining industry. He later constructed additional factories at Cressbrook, Bakewell, Wirksworth, Masson, Matlock, and Rochester, where he spun coarse counts of yarn for both the piece goods trade and foreign markets.[5]

Ownership of the workplace, machinery, motive power, and raw materials was typically concentrated in the hands of the factory master. In its initial stages of development, however, the new system of production allowed for considerable flexibility and experi-

---

lished between 1769 and 1802. These mills can be separated into two types on the basis of the technology employed. Throughout the north the small jenny workshop predominated. Corn mills, dwellings, even old barns and outbuildings of all descriptions were converted into jenny workshops where from eight to ten laborers operated hand carding machines, jennies, and twisting frames. Yet it was not among these factory masters that the major innovations in the textile industry occurred. Although the multistory, purpose-built factory was less common, it has attracted more historical attention. See Stanley D. Chapman, "Fixed Capital Formation in the British Cotton Industry, 1770–1815," *Economic History Review* 23 (August 1970):236–237, 256–266; and R. S. Fitton and A. P. Wadsworth, *The Strutts and the Arkwrights, 1758–1830: A Study of the Early Factory System* (Manchester, Eng., 1958), pp. 192–193; and Stanley D. Chapman, "The Peels in the Early English Cotton Industry," *Business History* 11 (July 1969):63.

4. John N. Merrill, "Arkwright of Cromford," *Industrial Archaeology* 10 (August 1973): 268–269; Stanley D. Chapman, "The Transition to the Factory System in the Midlands Cotton-Spinning Industry," *Economic History Review* 18 (December 1965): 529; Wadsworth and Mann, *Cotton Trade and Industrial Lancashire*, pp. 484–488.

5. Merrill, "Arkwright of Cromford," pp. 270–271; Fitton and Wadsworth, *Strutts and the Arkwrights*, p. 196.

mentation by the owner in the administration and operation of the factory, in the marketing of the product, in the motive power used, and in the recruitment of a labor force.[6]

Patterns of ownership and management that had traditionally been associated with mercantile pursuits and domestic industry were incorporated in the context of the new factory system. As a result of the Bubble Act of 1720, which for a century discouraged the formation of joint stock companies, most early industrial firms were organized as partnerships or, less frequently, as single proprietorships. In the early Cromford venture, for example, three men joined forces: Richard Arkwright, a barber by trade, and two hosiers, Samuel Need and Jedediah Strutt. Arkwright built the machines, recruited the labor, and managed the factory while his partners financed the operations. A slightly different arrangement was concluded by Samuel Greg, a merchant manufacturer who entered into a series of partnerships, taking on new associates as new techniques were introduced into the industry. Once established and successful, Greg gradually severed ties with his various partners and drew his sons into the business.[7]

For many early factory masters such as Greg, the closed family partnership was the preferred form of organization, and a few manufacturers were able to achieve considerable personal and familial control over their business ventures. Arkwright, for example, entrusted the Bakewell property to his son, Richard. Jedediah Strutt constructed five factories between 1779 and 1795; all were owned and managed by his family. While the closed family firm was the goal of many manufacturers, during the initial stage of industrialization it was difficult to achieve. Capital and technical expertise were the ingredients necessary for the successful opera-

6. Max Weber, *General Economic History*, trans. Frank H. Knight (New York, 1927), p. 302.

7. Greg formed his first partnership with John Massey in 1783 and constructed the Quarry Bank mill. Shortly after the factory was completed, Massey died, and Greg engaged Matthew Fawkner. Fawkner did not have the requisite knowledge to operate the mill successfully, and Greg turned to a celebrated engineer of Bolton and Watt, Peter Ewart. Ewart was a good choice, and in the decade following the formation of this partnership, the Quarry Bank mill was enlarged and improvements were initiated. Simultaneously Greg expanded operations, until by 1828 he owned factories in Lancaster, Caton, Bury, and Styal. See Mary B. Rose, "The Role of the Family in Providing Capital and Managerial Talent in Samuel Greg and Company, 1784–1840," *Business History* 19 (January 1977): 40–45.

tion of early factories, and few families possessed both. Most men were forced to look beyond the kinship network for partners.[8]

Marketing procedures exhibited less uniformity. Some of the largest manufacturers, including Robert Peel of Manchester and I. & R. Morley of Nottingham, opened warehouses in London and sold goods directly to retailers. Smaller manufacturers who could not afford to acquire storage facilities sold their yarn, stockings, and cloth through commission agents who operated in the major cities. While these commission agents charged a fee of approximately 5 percent for their services, they often proved invaluable to the small manufacturer by providing him with fashion advice, market information, and credit. A third group of manufacturers, including such giants in the industry as Jedediah Strutt, employed a variety of marketing schemes. In Strutt's case, agents acted on his behalf in the Leicester, London, Nottingham, and Manchester markets and received a commission for their services. At the same time, Strutt sold goods directly to customers from his factories and avoided the middleman's commission.[9]

The forms of motive power employed by early factory masters were even more diverse. In many cases locational factors and the availability of labor rather than personal preference dictated the motive power used. Initially factories were constructed in rural areas and used the power generated by swiftly flowing streams. But as many manufacturers moved from rural to urban locations, new forms of power were adopted. Richard Arkwright, who employed water wheels at Cromford and at several of his other factories, experimented with an atmospheric engine at his Manchester factory in 1792. Earlier, in 1786, James and John Robinson had installed a Bolton and Watt engine in their Nottingham mill, and several years later Peter Drinkwater demonstrated the success of similar engines in his Manchester factory. Despite the initial success of the steam

8. Ibid.; Chapman, "Peels in the Early English Cotton Industry," p. 64; Fitton and Wadsworth, *Strutts and the Arkwrights*, pp. 111, 169.

9. Orders sent directly to Strutt's firm were not filled unless accompanied by cash or by a written testimonial affirming the financial integrity of the customer. See Stanley D. Chapman, *The Cotton Industry in the Industrial Revolution* (London, 1972), p. 46; Fitton and Wadsworth, *Strutts and the Arkwrights*, pp. 271–274, 297–299; Geroge Unwin, *Samuel Oldknow and the Arkwrights: The Industrial Revolution at Stockport and Marple* (Manchester, Eng., 1924), pp. 56–57.

engine, it did not overtake immediately the more traditional source of power. Indeed, some manufacturers combined the two, using water power for much of the year and reserving their engines for emergencies. Others used water wheels in some of their mills and steam engines in others.[10]

Labor practices also varied considerably. Although all manufacturers employed children, three different systems operated in the British textile industry from the mid-eighteenth century to its close: apprenticeship labor, contract wage labor, and family labor. From the outset manufacturers found it impossible to hire an adult labor force, and so they turned to children.[11] Because factory work required little physical strength or skill, it appeared well suited to boys and girls. In Britain children traditionally had been a major component of the labor force. In agriculture young children labored beside their parents from dawn to dusk; in the mines they worked as trappers, opening and closing ventilation doors; in the

10. Arkwright is the obvious example here, but Peter Drinkwater installed a steam engine in his Manchester factory and continued to spin by water at his Northwich (Cheshire) mill. See Stanley D. Chapman, "The Cost of Power in the Industrial Revolution in Britain: The Case of the Textile Industry," *Midland History* 1 (1971): 1–23; Jennifer Tann, "Richard Arkwright and Technology," *History* 58 (February 1973): 29–44; Andrew Ure, *Philosophy of Manufacturers, or an Exposition of the Scientific, Moral, and Commercial Economy of the Factory System of Great Britain* (London, 1835), p. 344; J. Britton and E. W. Brayley, *The Beauties of England and Wales*, quoted in H. R. Johnson and A. W. Skempton, "William Strutt's Cotton Mills, 1793–1812," *Newcomen Society for the Study of the History of Engineering and Technology* 30 (1960): 190; Chapman, "Fixed Capital Formation," pp. 239–241; Robert Owen, *The Life of Robert Owen by Himself* (New York, 1920), p. 53.

11. Agricultural workers were accustomed to seasonal, task-oriented labor; during the spring and summer months, they usually tramped over the countryside in search of work, sometimes laboring twelve to fourteen hours daily, but this brisk activity was soon followed by a period of relaxation. Those in other occupations were also accustomed to task-oriented, not time-oriented, labor. Weavers, framework knitters, shoemakers, and others engaged in cottage industries set their own pace, constantly interrupted their activities, and worked generally on their own rather than under the supervision of others. The factory system with its emphasis on regularity, discipline, and punctuality was repugnant to many of these people. See Wadsworth and Mann, *Cotton Trade and Industrial Lancashire*, p. 406; Michael Anderson, *Family Structure in Nineteenth-Century Lancashire* (Cambridge, Eng., 1971), pp. 137–138; Chapman, *Cotton Industry in the Industrial Revolution*, p. 53; Pollard, *Genesis of Modern Management*, pp. 162–163; Harold Perkin, *Origins of Modern English Society, 1780–1880* (London, 1967), p. 159; Edward P. Thompson, "Time, Work-Discipline, and Industrial Capitalism," *Past and Present* 38 (December 1967): 61; Evelyn G. Nelson, "The Putting-Out System in the English Framework Knitting Industry," *Journal of Economic and Business History* 2 (May 1930): 472–473.

pillow lace industry boys and girls as young as three and four were employed to handle bobbins.[12]

The rise of the factory system extended the use of child labor. Manufacturers found that children as young as three could collect cotton waste that accumulated near machinery and that children of seven and upwards could be employed to operate the simple carding and spinning equipment. Although very young children were used in a few factories, most manufacturers refused to employ children under seven years of age.[13]

Among the first children introduced to factory work were pauper apprentices. In the eighteenth century parents who could not provide adequately for their children were encouraged to place them in workhouses so that they might "be bred up to labour, principles of virtue implanted in them at an early age, and laziness be discouraged."[14] Local officials had the authority to bind out the youngsters for a specified number of years. Several of the domestic textile trades, including the silk industry, the pillow lace industry, and the framework knitting industry, employed this type of labor. In the pillow lace industry, for example, youngsters worked twelve to fourteen hours daily at the simple and monotonous task of drawing lace. The children received no training in this industry. After completing their apprenticeship, these young people still had to learn a trade.[15] This form of pauper apprenticeship was transferred to the new textile factories.

From London, Liverpool, Cheshire, Newcastle-under-Lyme, indigent or abandoned or orphaned children drawn from local workhouses, poorhouses, foundling homes, and schools of industry were transported to the northern mills of Samuel Greg, Robert Peel, John Smally, and Samuel Oldknow to serve an "apprentice-

12. Ivy Pinchbeck and Margaret Hewitt, *Children in English Society*, 2 vols. (London: 1969–1973), 2:390–393 and 2:395–396.
13. Ibid., 2:354 and 2:404; see also Fitton and Wadsworth, *Strutts and the Arkwrights*, pp. 226–227.
14. R. K. Webb, *British Working-Class Reader, 1790–1848: Literacy and Social Tension* (London, 1955), p. 113.
15. Pinchbeck and Hewitt, *Children in English Society*, 2:391 and 2:396; see also M. G. Jones, *The Charity School Movement: A Study of Eighteenth-Century Puritanism in Action* (London, 1938), pp. 28–31; Wadsworth and Mann, *Cotton Trade and Industrial Lancashire*, p. 408; Pollard, *Genesis of Modern Management*, p. 164.

ship." Richard Arkwright also experimented with this form of labor at his Cressbrook mill in 1799.[16] These children, who ranged in age from seven to fourteen, provided the bulk of the textile labor force. Boys and girls worked twelve hours daily, six days a week, and on Sunday they attended church in the morning and school in the afternoon. Accommodation, food, clothing, and recreational, religious, and educational activities were supplied by the factory master. This appeared a convenient, economical way to obtain and maintain a labor force.

Although many of the manufacturers who employed these children treated them well, the system was open to considerable abuse. The "dark satanic mills" depicted in novels and textbooks alike can be traced to some of the factories employing pauper apprentices.[17] While small-scale factory masters who operated on a shoestring were considered the worst offenders, abuse occurred even among the largest and the best-known employers of pauper apprentices. Robert Peel, who employed approximately one thousand young apprentices in his factories, described the conditions of his work force:

> I was struck with the uniform appearance of bad health, and in many cases, stinted growth of the children; the hours of labour were regulated by the interest of the overseer, whose remuneration depending on the quality of the work done, he was often induced to make the poor child work excessive hours, and to stop their complaints by trifling bribes.[18]

Peel saw few alternatives to the use of pauper apprentices: "I believe in country places there was no opportunity of getting young

16. Pinchbeck and Hewitt, *Children in English Society*, 1:354, 2:391, 2:396; Jones, *Charity School Movement*, pp. 28–31; Wadsworth and Mann, *Cotton Trade and Industrial Lancashire*, p. 408; Pollard, *Genesis of Modern Management*, p. 164; M. H. Mackenzie, "Cressbrook and Litton Mills, 1779–1835: Part 1" *Derbyshire Archaeological Journal* 88 (1968): 4–5; Merrill, "Arkwright of Cromford," p. 273.

17. Chapman, *Cotton Industry*, p. 56; Frances Collier, *Family Economy of the Working Classes in the Cotton Industry, 1784–1833* (Manchester, 1964), p. 45; Ure, *Philosophy of Manufacturers*, p. 347; Great Britain, Parliament, *Sessional Papers* (Lords), 1819, 108, "An Account of the Cotton and Woollen Mills and Factories, 1803–1818," March 8, 1819, April 28, 1819, and August 11, 1807; Unwin, *Samuel Oldknow and the Arkwrights*, p. 173.

18. Great Britain, Parliament, *Parlimentary Papers* (Commons) 1816, 3, "Select Committee on the State of the Children Employed in the Manufactories of the United Kingdom," p. 132.

labour but from large towns abounding with population, and those I understood to be uniformly apprentices."[19] From the beginning apprentices were found throughout the textile industry. Only during the closing years of the eighteenth century did this system come under investigation and eventually fall into disrepute. In 1802 the Health and Morals of Apprentices Act was passed; its purpose was to set minimum standards for apprentices working in cotton factories by eliminating night work, limiting to twelve the number of hours a child could work, and forcing employers to provide both religious and educational instruction to their charges.[20] The law, however, was easily contravened.

While the outline and implications of pauper apprenticeship as a labor form are clear, less is known about female and child contract wage labor. To run his Cromford mill, Richard Arkwright initially hired the wives and children of local lead miners. Expected to board themselves, these workers agreed to work for a specific wage and for a predetermined length of time. Boys and girls, some as young as seven years, served as machine operators. Approximately four-fifths of these children were boys. In 1777 Arkwright employed two hundred people, and within two decades about six hundred additional hands entered his factories. He worked his labor force around the clock; mainly boys were employed on the night shift. Arkwright, however, could not secure all the labor required from within the district, and he advertised in Manchester and Nottingham for workers to be hired on contract.[21]

Other manufacturers were slow to adopt contract labor. There is evidence, however, that at the turn of the century Samuel Greg, Robert Peel, and Samuel Oldknow, among others, began to supplement and then to replace their pauper apprentices with contract workers. This transition took decades to complete, and for many years, contract workers could be found laboring side by side with

19. Ibid., p. 140.
20. Mackenzie, "Cressbrook and Litton Mills," pp. 10–11; Arthur Redford, *Labour Migration in England, 1800–1850* (Manchester, 1964 [1926]), pp. 26–33.
21. Stanley D. Chapman, *Early Factory Masters: Transition to the Factory System in the Midlands Textile Industry* (Newton Abbot, Eng., 1967), p. 168; Merrill, "Arkwright of Cromford," pp. 270–275; James Pilkington, *A View of the Present State of Derbyshire*, 2 vols. (Derby, 1789), 1:301; Fitton and Wadsworth, *Strutts and the Arkwrights*, p. 105; and Parliament, "Select Committee on the State of the Children Employed in the Manufactories of the United Kingdom" (1816), pp. 277–280.

pauper apprentices. The pace of this transition and the various forms it took could be observed at the Styal mill of Samuel Greg. Around 1783 Greg constructed a mill at Quarry Bank for the manufacture of cotton warps, and there he employed approximately 80 apprentices and 183 contract workers. To secure free—that is, nonapprenticed—laborers, Samuel Greg directly contacted indigent families and arranged with them for the services of their sons and daughters. In exchange for their labor, Greg provided the children with accommodation, food, and a small weekly wage. Through poor-law commissioners, Greg later engaged entire families and persuaded them to move to Styal. For them special housing and other facilities had to be introduced. Pauper apprentices and contract labor worked together in the factory for several decades. It was not until 1847 that the Greg family abandoned apprentices altogether and operated the firm with a voluntary labor force.[22]

Family labor was also employed by early factory masters. Less is known about this labor system than about the other two. Under this system, all members of the family worked for the manufacturer in some capacity. Children and adolescent boys and girls worked in the mills; married women often took in piecework such as picking—a hand process performed in the home and not subject to factory discipline. Male householders usually avoided the factory and worked building bridges, roads, canals, and houses, operating kilns, or transporting goods.[23]

Family labor appeared to be the last choice of early factory masters. Many probably turned to this system after the supply of local children had been exhausted and after pauper apprentices had been employed and were found to be inefficient, expensive, or troublesome. Several factors support this conclusion. Respectable householders equated the factory with the workhouse and factory employment with a degraded type of labor, one appropriate for

---

22. Collier, *Family Economy of the Working Classes*, pp. 39–45; Rose, "Role of the Family," p. 40.

23. Ure, *Philosophy of Manufacturers*, p. 289; Collier, *Family Economy of the Working Classes*, p. 17; Perkin, *Origins of Modern English Society*, p. 151; Anderson, *Family Structure in Nineteenth Century Lancashire*, pp. 71–74; Chapman, *Cotton Industry*, pp. 20, 58; Unwin, *Samuel Oldknow and the Arkwrights*, pp. 142, 169, 204–207; 222–223, 215–221.

indigent, sick, or disreputable people. Unless destitute or otherwise desperate, adults tried to avoid the new factory system. Furthermore, the economic situation did not force respectable families to enter the mills. Many householders had opportunities that were both socially more acceptable and financially more rewarding than factory employment. Agricultural workers could earn more than factory operatives; hand-loom weavers experienced relatively good times in these early years; and work was available for men in transportation, construction, coal mining, and other industries. For unskilled hands, domestic industry workers, and agricultural laborers, the factory held little attraction.[24]

Evidence drawn from management's perspective also clarifies the reluctance of factory masters to introduce family labor. This form of labor was expensive. The employment of householders in tasks not associated with yarn production represented a drain on capital resources which many early factory masters could ill afford. The cost argument can be extended, for the employment of family labor necessitated the building of homes, stores, and other facilities—in fact, the designing of entire communities to meet the needs of householders and their families. For many factory masters this represented a prohibitive expense. Stanley D. Chapman estimated that Richard Arkwright at Cromford and Jedediah Strutt at Belper spent approximately £60 for the construction of each single-family dwelling, an outlay of capital well beyond the means of most early factory masters. Even a manufacturer as wealthy as Strutt did not construct worker housing until well after his first factory was completed.[25] An argument can be made that family labor and the large factory towns synonymous with that form of labor belong to a later industrial era.

Once workers were recruited and provided with accommodation, their old habits had to be replaced with a new work discipline.

24. Redford, *Labour Migration in England*, p. 22; Perkin, *Origins of Modern English Society*, p. 128.
25. Stanley D. Chapman, "Workers' Housing in the Cotton Factory Colonies, 1770–1850," *Textile History* 7 (1976):119–122; see also L. D. W. Smith, "Textile Factory Settlements in the Early Industrial Revolution, with Particular Reference to Housing Owned by Cotton Spinners in the Water Power Phase of Industrial Production," Ph.D. thesis, University of Aston in Birmingham, 1976, pp. 51, 77; Fitton and Wadsworth, *Strutts and the Arkwrights*, p. 102.

Operatives had to adjust to the time schedule of the factory and to the speed and regularity of machinery. Again, several approaches were taken to solve these problems. Through corporal punishment, fines, or the threat of dismissal, some factory masters sought to create a docile, disciplined, and steady labor force.[26] Others emphasized self-control. Factory masters designed programs in which workers were made to feel guilty for deviant conduct. Although controversy surrounds the origin of one of the most popular of these schemes, the Sunday school, most scholars consider Robert Raikes to be its founder. Raikes organized schools where, on Sundays, young children supposedly were taught reading and the church catechism and were generally kept busy.[27] Started in 1781, the Sunday school movement spread quickly throughout Britain and became the special project of evangelical clergymen, provincial philanthropists, and manufacturers alike. Jedediah Strutt adopted the Sunday school as early as 1784, and soon 120 scholars attended classes. With Strutt, attendance was mandatory. Through sermons, songs, prayers, and recitation exercises, children were taught lessons on punctuality, order, obedience, deference, and other values thought necessary for the development of a tractable, industrious labor force.[28]

The Strutt family did not rely solely on internal forms of self-

26. Great Britain, Parliament, *Parliamentary Papers* (Commons) August 8, 1832, "Bill to Regulate the Labour of Children in the Mills and Factories of the United Kingdom," pp. 5–13 and 17–26. W. Cooper testified that he was often strapped by the overseer when he made mistakes or when he flagged in his work. Beatings or cuffings by the overseer were common throughout the industry. For minor infractions of factory discipline such as singing, swearing, and throwing objects, fines sometimes equaling a day's pay were deducted from workers' wages. Yet the greatest penalty was reserved for those hands who tried to organize their fellow workers against the factory master; they were dismissed and blacklisted. See Pollard, *Genesis of Modern Management*, pp. 184, 187–188.

27. Walter Thomas Laqueur, *Religion and Respectability: Sunday Schools and Working-Class Culture, 1780–1850* (New Haven, 1976), pp. 21–23. Jones, *Charity School Movement*, pp. 142–144, 146; Pinchbeck and Hewitt, *Children in English Society*, 2:293–296.

28. *Derby Mercury*, August 25, 1785, quoted in Fitton and Wadsworth, *Strutts and the Arkwrights*, pp. 102–103; see also A. P. Wadsworth, "The First Manchester Sunday Schools," John Rylands Library, Manchester, *Bulletin* 33, p. 300. Jones, *Charity School Movement*, pp. 3–4; Webb, *British Working-Class Reader, 1790–1848*, p. 15; J. H. Plumb, "Children in Eighteenth-Century England," *Past and Present*, May 1975, p. 83; Pollard, *Genesis of Modern Management*, p. 193. See also Laqueur, *Religion and Respectability*, pp. 187–240.

control, however, and its Sunday schools were designed to supplement, not supplant, fines and other forms of punishment. Absence without leave, theft, destruction of factory property, failure to complete assigned tasks, misconduct both inside and outside working hours, all carried fines.[29]

In this booming, diversified industry, Samuel Slater received his training. Born in Belper, Slater was apprenticed to the local factory master, Jedediah Strutt, in 1783. For six years he learned the "mystery" of the cotton textile industry and became part of a growing class of middle-level managers who emerged toward the close of the eighteenth century. These men, who stood below the proprietor but above the overseers and second hands, were trained to assume many of the responsibilities of day-to-day operations. While their duties varied from one factory to another, they had oversight of the mill and probably supervised the blending of cotton; sometimes they constructed and repaired machinery, recruited workers, disciplined hands, paid wages, and maintained good order in and around the factory.[30]

Slater had a thorough knowledge of these duties and of the British textile industry when he immigrated to the United States in

29. The Strutts believed that both the discipline imposed by the factory and the values taught in the Sunday school had positive effects on the children and on the community in general. In testimony before the Select Committee in 1816, George Strutt boasted: "It is well known in this neighborhood, that before the establishment of these works the inhabitants were notorious for vice and immorality, and many of the children were maintained by begging; now their industry, compared with the neighboring villages, where no manufacturers are established, is very conspicuous." See Parliament, "Select Committee on the State of the Children Employed in the Manufactories of the United Kingdom" (1816), pp. 217–218. See also Fitton and Wadsworth, *Strutts and the Arkwrights*, pp. 233–236. Fines provided an additional disciplinary tool, but Strutt never resorted to corporal punishment. He adopted a humane, somewhat paternalistic attitude towrd workers, their discipline, and their welfare. And his sons shared his attitudes. "Of men of great practical measures," wrote Robert Owen of New Lanark, "were Mr. William Strutt, . . . and his brother Joseph, two men whose talents in various ways and whose truly benevolent dispositions have seldom been equalled." This is high praise indeed coming from one of the most highly regarded factory masters of the era. And Strutts' biographers would argue: "The idealized community which Robert Owen thought he had invented at New Lanark was not much different from those at Cromford and Belper that had preceded it." See Owen, *Life of Robert Owen*, pp. 291–292; Fitton and Wadsworth, *Strutts and the Arkwrights*, p. 98. See also Bendix, *Work and Authority in Industry*, p. 47.

30. Owen, *Life of Robert Owen*, p. 39; Pollard, *Genesis of Modern Management*, pp. 104–192.

1789. His British experience provided him with set of options that would guide him in the construction of a factory and in the recruitment and discipline of a labor force. The American factory system that emerged under his guidance represented an adaptation, not a carbon copy, of his British experience.

[2]

Almy and Brown, American Merchants

The transfer of any process, institution, or idea from one society to another, even among societies that share common values and customs, can never be complete. Some compromise with the realities of the host community inevitably must take place, and this was especially true for so novel a process as the factory system. Although in America Samuel Slater tried to introduce British programs on the ownership and management of a factory and on the recruitment of labor, he met considerable resistance from the local population. Merchants, laborers, and farmers had definite opinions regarding this new process and its implementation. Slater encountered opposition from his American partners, William Almy and Smith Brown, as well as from his workers. The demands of the factory system clashed with their traditional ideas on business procedures and family responsibilities. Out of a decade of experimentation by and struggle among Slater, his partners, and his workers, an American fctory system emerged. Although it reflected its British heritage, the factory system that came to carry Slater's name from Maine to Pennsylvania was a product of its American environment.

In 1789, when Slater arrived in America, the textile industry was still in the household and early cottage stage of development. Flax and wool continued to be carded and spun in the home for personal consumption; some of the yarn was also taken to local weavers, who worked it up into low-quality sheeting and shirting. The

[47]

weaving workshops that had been established were small; they housed between two and four looms and provided employment for a master weaver and several apprentices. To be sure, there were a few large workshops, such as the one operated by Peter Colt of Paterson, New Jersey. Colt and other merchant manufacturers wove large quantities of goods for sale to the public. In Colt's workshop, a head workman supervised eight hand-loom weavers while Colt devoted most of his time to marketing the finished hosiery.[1] Larger workshops, with ten, twenty, even fifty looms, also could be found in the vicinity.[2] While there was a market for mass-produced piece goods, the absence of a sufficient quantity and a definable quality of yarn prevented the expansion of American shops along British lines. In Britain the rise of spinning factories had broken this bottleneck and had allowed for the rapid expansion of the weaving industry. Merchants, professional men, politicians, and others believed that the establishment of spinning mills in the United States would guarantee comparable results.[3]

Gaining access to the new technology proved a major obstacle, as designs and instructions for operating carding and spinning machines were known only to the artisans who built the equipment, and they were prohibited from leaving Britain. Anyone who encouraged them to emigrate was subject to a fine, and anyone who tried to export the machinery faced loss of his own equipment,

1. John G. Palfrey, *History of New England: During the Stuart Dynasty*, 2 vols. (Boston, 1860), 2:53; Eliza Philbrick, "Spinning in the Olden Time," *Essex Antiquarian* 1 (June 1897):88; Providence, R.I., Early Records of the Town of Providence, Rhode Island, 16:1, 26, 88, 109, 137, 189, 217, 236, 323, 385, 463, quoted in Rolla M. Tryon, *Household Manufactures in the United States, 1640–1860* (Chicago, 1917), p. 84; William B. Weeden, *Early Rhode Island: A Social History of the People* (New York, 1919), p. 120; Victor S. Clark, *History of Manufactures in the United States*, 3 vols. (New York, 1949 [1929]), 1:189–190; Louis Bard to Moses Brown, Norwich, April 2, 1792, Almy and Brown Papers, Samuel Slater Production Reports and Correspondence, Rhode Island Historical Society, Providence (hereafter referred to as Almy and Brown MSS); David John Jeremy, ed., *Henry Wansey and His American Journal, 1794* (Philadelphia, 1970), p. 125; and Almy and Brown MSS, Peter Colt to Almy and Brown, Hartford, May 22, 1793.
2. Clark, *History of Manufacturers*, 1:189–190.
3. Kenneth F. Mailloux, "Boston Manufacturing Company: Its Origins," *Textile History Review* 4 (October 1963):157–158; Arthur H. Cole, *Industrial and Commercial Correspondence of Alexander Hamilton Anticipating His Report on Manufactures* (Chicago, 1928), pp. 248–259; Robert W. Lovett, "The Beverly Cotton Manufactory, or Some New Light on an Early Cotton Mill," *Bulletin of the Business Historical Society* 26 (December 1952):220–224.

imprisonment, and a fine.[4] To surmount these difficulties, Americans tried a number of schemes, some of them ingenious and others quite foolish. A few enterprising men met every incoming vessel and searched among the passengers for machine builders. Others traveled to Britain and visited the textile districts, hoping to entice employees away or somehow to learn enough about the machinery to reproduce the designs themselves. Thomas Digges, the disreputable son of a wealthy Maryland family, gained considerable notoriety for his efforts in this area.[5] Most of those concerned with the acquisition of technology, however, pursued a less risky course. Through advertisements placed in local British newspapers, and through personal correspondence with known artisans, they induced British mechanics to emigrate. Once in New York, Pennsylvania, or Massachusetts, the mechanics were contracted by individuals, societies, or state officials to reproduce Arkwright designs.[6]

Moses Brown was one such enterprising individual who realized the potential market for this yarn. Brown was related to one of the leading mercantile families in Rhode Island. In the 1750s Obadiah Brown, his uncle, had accumulated a sizable fortune through interests in a chocolate mill, an iron furnace, and a spermaceti candle business. As his uncle's apprentice, Moses Brown had kept the accounts and had held responsibility for coordinating an extensive trading network in both the iron and the candle enterprises. In 1762 the businesses passed to Moses Brown and his three brothers. Reorganized as Nicholas Brown and Company, the new firm traded candles and pig iron for British manufactured goods and sold these products to shopkeepers throughout Connecticut, Rhode Island, and Massachusetts.[7] A decade later Moses Brown retired, became

---

4. David John Jeremy, "Damming the Flood: British Government Efforts to Check the Outflow of Technicians and Machinery, 1780–1843," *Business History Review* 51 (Spring 1977):1–2.

5. Carroll W. Pursell, Jr., "Thomas Digges and William Pearce: An Example of the Transit of Technology," *William and Mary Quarterly* 21 (October 1964):552–557; Lynn Hudson Parsons, "The Mysterious Mr. Digges," *William and Mary Quarterly* 22 (July 1965):486–492.

6. Paul E. Rivard, "Textile Experiments in Rhode Island, 1788–1789," *Rhode Island History* 33 (May 1974):36–37; James Lawson Conrad, Jr., "The Evolution of Industrial Capitalism in Rhode Island, 1790–1830: Almy, the Browns, and the Slaters," Ph.D. dissertation, University of Connecticut, 1973, p. 24.

7. Almy and Brown MSS, Almy and Brown to (no name given), October 7, 1791; Moses Brown to Samuel Slater, December 10, 1789, quoted in William R. Bagnall,

a Quaker, and devoted his energies to social issues. Because of financial pressure and a desire to see his son-in-law, William Almy, established firmly in a business, Brown reemerged from his self-imposed retreat in 1788. His new interests centered on textile manufacturing rather than on iron or candle production. He purchased a complement of spinning and weaving equipment including crude spinning frames, hand cards, spinning jennies, stocking frames, and hand looms. Shortly thereafter he formed the partnership of Almy and Brown with William Almy and Smith Brown, his cousin. They set up as merchant manufacturers in the textile trade and put out wool and flax to local pickers and spinners; they also purchased linen yarn for warps and sold the cloth manufactured by weavers in their workshop. Managerial operations were left in the hands of Almy and Smith Brown while Moses Brown provided guidance and encouragement from the wings. Moses Brown realized quickly, however, that the business could not prosper under these conditions, and he wanted to introduce mechanical spining, then widely practiced in Britain.[8]

Samuel Slater was employed at a jenny workshop, the New York Manufacturing Company, when rumors reached him that Brown, who years earlier had acquired spinning equipment, was searching for a mechanic who could set it into operation. Desiring to leave his position, Slater contacted the merchant, outlined his own qualifications, and requested further information on the experiment. After considerable discussion between the two men, it became clear that an agreement could not be effected easily. While Brown merely wanted Slater to repair and operate the equipment, Slater demanded a share in the business.[9] Men with practical knowledge of textile machinery and factory operations were at a premium in America, and he knew it. Certainly he could command more for his experience than the inducement offered by Brown.

Slater made a counteroffer. He would construct new equipment

Textile Industries of the United States (Cambridge, Mass., 1893), p. 152; see also John W. Haley, *Lower Blackstone River Valley* (Pawtucket, 1937), pp. 50–51; Hedges, *Browns of Providence Plantations*, pp. 13–14, 184–187; "Beginnings of the Industrial Revolution in America," pp. 98–101.

8. Conrad, "Evolution of Industrial Capitalism in Rhode Island," pp. 41–45.

9. Almy and Brown MSS, Almy and Brown to (no name given), October 7, 1791; Moses Brown to Samuel Slater, December 10, 1789, quoted in Bagnall, *Textile Industries of the United States*, p. 152.

based on British models for a share in the new enterprise. After considerable negotiation, William Almy, Smith Brown, and Samuel Slater formed a partnership whereby Almy and Brown would underwrite the venture and Samuel Slater would organize and oversee the production process. Two years after the partnership agreement was signed, Smith Brown retired to Massachusetts and was replaced by Obadiah Brown. Under the agreement, Slater promised to build the equipment, to construct a mill, to supervise production, and to pay one-half the expenses of the project. His partners agreed to purchase raw materials, to sell finished goods, to pay one-half the cost for this operation, and to advance Slater sufficient funds to fulfill his financial part of the bargain.[10] From the start each of the partners held a different perspective on his role in the new venture. Almy and Brown viewed the factory as a component part of their larger business and not as a separate entity. Rather than obtain spun yarn from outworkers in the vicinity for their weaving workshop, they hired Slater to provide them with this product. Slater, however, did not share this limited vision of his position. The production process he set up had little in common with the domestic system familiar to Almy and Brown. Fully mechanized, using water rather than human energy as a motive power and operated by a corps of people dependent primarily on wages for their livelihood, Slater's "factory" was unique; nothing similar existed anywhere in the United States. Yet Slater could not convince Almy and Brown that he alone knew best how to run this factory, that he was a competent, trustworthy partner, and that he deserved an equal share in the business.[11]

THE YARN-SPINNING MILL

In a clothier's shop in Pawtucket, Slater installed carding machines, water frames, and a carding and roving machine. Children

10. Harold Hutcheson, *Tench Coxe: A Study in Americn Economic Development* (Baltimore, 1938), p. 161; Almy and Brown MSS, Moses Brown to John Dexter, July 22, 1791; and George S. White, *Memoir of Samuel Slater, the Father of American Manufactures Connected with a History of the Rise and Progress of the Cotton Manufacture in England and America with Remarks on the Moral Influence of Manufactories in the United States* (Philadelphia, 1836), pp. 74–75.
11. Kulik, "Beginnings of the Industrial Revolution in America," pp. 148–162.

between seven and twelve years of age were hired to operate the new equipment. Within a few years the machinery was moved to a specially built "factory"; later, additional equipment was constructed and more laborers were hired. By the turn of the century, more than one hundred people worked in the yarn-spinning mill.[12]

Although the Pawtucket factory proved an unqualified success, the working relationship between Slater and his Providence partners was never harmonious. Almy and Brown saw Samuel Slater as an employee, a manager of the yarn-spinning component of their business, a man to keep in check. At the outset they determined to restrict his role and influence not only in the overall operation of their textile business but also in the administration of the Pawtucket side of the venture. From their Providence store they directed every phase of the enterprise, including the organization and management of the Pawtucket mill. The influence and control they exercised centered on their monopoly of the finances. Slater remained completely dependent on Almy and Brown for supplies and funds. Their inability or unwillingness to provide him with hardware, foodstuffs, and cash occasioned considerable delay in factory operations, diminished Slater's status and influence with workers, and caused constant friction between the partners.

From their Providence retail store, Almy and Brown handled all financial transactions. Supplies for Slater's operations—primarily raw cotton—consumed their time and money. From importers and domestic merchants they purchased West Indian cotton, which was considered cleaner and had a longer staple than the American variety. In the United States, unripe cotton was picked and intermixed with ripe cotton, and plant membranes, pods, seeds, dirt, and other foreign matter were packed along with the cotton. According to Moses Brown, "the present production in the mixed manner in which it is Brought to Market does not answer good purpose." He preferred "to have his supply from the Westindies Under the Discouragement of the Impost, rather than Work our

---

12. White, *Memoir of Samuel Slater*, p. 99; Ware, *Early New England Cotton Manufacture*, p. 23; William B. Browne, *Genealogy of the Jenks Family of America* (Concord, 1952), pp. 109, 211; Brendan F. Gilbane, "A Social History of Samuel Slater's Pawtucket, 1790–1830," Ph.D. dissertation, Boston University Graduate School, 1969, pp. 75–84.

Own production."[13] By the mid-1790s, cotton from the southern states had improved in both quality and packaging, and the partners switched to the home-grown product and purchased large quantities of Georgia cotton from Providence and Newport merchants. Costs, however, were high: in 1796 they paid $100.84 for a 242-pound bale of local cotton.[14] The cotton was shipped directly to the Providence store. There Almy and Brown divided the bales into small parcels, supervised the initial cleaning of the cotton, and then sent the small packages one by one to Slater.[15]

By 1795 expenses for raw cotton represented approximately two-thirds of the operating costs for the mill. Other expenditures were for building materials, hardware, food products, clothing, shoes, wages, marketing costs, and other incidentals.[16] Not allowed to purchase supplies on his own account, Slater submitted lists of supplies to Almy and Brown. "The mill is now destitute of the following articles cotton to pick, corn, rye, coffee, tea, molasses and flour therefore if you have a part or all or can procure the above said articles, you will please to send them up as soon as conv. [convenient]," read one such request.[17] This memorandum included supplies for the factory and commodities for laborers. Almy and Brown paid workers in both cash and store goods. While many hands accepted part of their wage payment in corn, rye, flour, molasses, rice, sugar, and other commodities, others preferred clothing, boots, shoes, or personal articles. All goods came from Almy and Brown's Providence retail store.[18]

Supply channels often broke down. The Providence merchants

13. Almy and Brown MSS, Moses Brown to John Dexter, July 1791.
14. Ibid., William Almy to Samuel Slater, September 18, 1795, quoted in White, *Memoir of Samuel Slater*, p. 189; Almy and Brown MSS, Thomas Arnold to Almy and Brown, November 25, 1796.
15. Conrad, "Evolution of Industrial Capitalism in Rhode Island," pp. 71, 77.
16. Ware, *Early New England Cotton Manufacture*, p. 124.
17. Almy and Brown MSS, Samuel Slater to Almy and Brown, April 25, 1801.
18. Aside from special tools, screws, and replacement parts for the machinery which could be found only in New York, Boston, or Philadelphia, most of the hardware, food, and other supplies required by Samuel Slater were purchased within the area. See Almy and Brown MSS, Materials Purchase, January 29, April 19, and May 11, 1796; Richard Harding to Almy and Brown, May 3, 1796; Samuel Slater to Almy and Brown, September 20, 1796, and July 30, 1806; Lawton to Almy and Brown, April 29 and September 13, 1809. See also Spinning Mills, vol. 1, Spinning Mills to Almy and Brown, February 14 and April 12, 1800, Samuel Slater Collection, Baker Library, Harvard University (hereafter referred to as Slater MS).

did not understand that regularity of supply was essential to the efficient operation of the mill. In 1794, 1795, and 1796 Slater ran short of raw cotton and had to threaten to shut down the mill unless he was supplied immediately. In 1796, for example, he wrote to his partners: "The Machinery is now principally stopped for want of cotton wool." One week later, he again wrote: "Please send some fleece cotton . . . if not I must unavoidably stop the mill after this week."[19] Other supplies, including shovels, brooms, and supplies for the factory hands, were seldom delivered on schedule. In desperation Slater once wrote: "Brushes much wanted!!! none to sweep the mill with."[20] This was no laughing matter. The lint and dust that accumulated in the mill posed a dangerous fire hazard. A flame from any one of the open lamps or a spark from the machinery could easily ignite the combustible materials and destroy the factory.[21]

Supply deficiencies caused considerable friction, but they produced far fewer arguments between the partners than the treatment accorded labor. Wage payment was the issue. Almy and Brown failed to understand that the survival of scores of people in their factory rested on the prompt and regular payment of wages in cash or kind. Parents wanted to know precisely how much each child earned per week, when they could expect payment, what form payment would take (cash or kind), and where they could collect the wages or the goods. In these matters Almy and Brown equivocated. Sometimes they sent Samuel Slater sufficient funds and supplies to pay off the help, and at other times they forced house-

19. Almy and Brown MSS, Samuel Slater to Almy and Brown, March 20, 1794; June 18, 1795; February 12 and 19, 1796; April 25, 1801.
20. Ibid., Samuel Slater to Almy and Brown, April 25, 1801; October 9, 1811; Lawton to Almy and Brown, December 2, 1808.
21. This did, in fact, occur. Several years later Slater described a potentially disastrous scene:
About 45 minutes past ten last night, the disagreeable noise of a cry of fire, fire was again heard thro the streets in this place which was in our Spinning Room when the people first got to the Factory, the top frame over one of the spinning frames blazed freely, but Joshua Vaughn and a few others broke the sash in and with the help of Pales only, put the fire out before the Fire Engine got there. I do think that the present alarming circumstance requires our immediate attention to the adoption of some method or other.
See Almy and Brown MS, Samuel Slater to Almy and Brown, October 9, 1811; Lawton to Almy and Brown, December 2, 1808.

holders to travel to Providence to collect the wage payments from them at their store. Furthermore, they failed to settle cash accounts with any regularity. Some hands received cash payments each month, while others had to wait six or even eight months for a settlement.[22] Disagreements over these issues drove a wedge between the partners: Almy and Brown refused to believe that these were concerns that demanded immediate attention.[23]

The credit system with which Almy and Brown were familiar as merchants was haphazard at best and explained part of their difficulty in honoring their commitments promptly. They sold goods on six- to nine-month credit and often found it impossible to collect debts. Absence of ready cash, poor communication routes, and, most important, a nonchalant attitude toward payment was part of the accepted code of mercantile operations.[24] Furthermore, Almy, Brown, and Slater were in debt. Initially they borrowed over $10,000 from Moses Brown, and this indebtedness increased when they borrowed additional funds to construct a new factory. During the early years the firm struggled to maintain production and to expand markets. As late as 1796 the debt owed Brown alone totaled between $15,000 and $20,000.[25] The firm possessed no financial reserves, and Almy and Brown were hard pressed to supply Slater with cash even when they wanted to fulfill their obligations to workers.

Economic circumstances peculiar to the era explain only part of Almy and Brown's actions during the first decade of operation. Throughout the era, they harassed Slater, trying to reduce his influence with the labor force and to limit his discretion. The supply of raw materials was a case in point. Almy and Brown could have had the raw cotton sent from Providence and Newport wholesale firms directly to the factory rather than to their Providence store. Or they could have supplied Slater with larger lots of cotton rather than doling it out in small portions and forcing him to plead for materials. While this method of supply occasioned delay, it allowed

22. Ibid., Samuel Slater to Almy and Brown, October 17, 1793; June 2, 1795.
23. Ibid., June 2 and September 25, 1795; January 7, October 4, and November 10, 1796.
24. Chandler, *Visible Hand*, pp. 15–19, 22, 36–40.
25. Conrad, "Evolution of Industrial Capitalism in Rhode Island," pp. 96–97.

Almy and Brown to dictate the pace of factory operations. Authority over the work force was also divided. Slater supervised workers, but Almy and Brown set wages, determined the mode of payment, established the settlement schedule, and served as a court of appeal when disputes erupted between Slater and the workers. In one case, the father of several hands failed to accept Slater's decision and threatened to press his demands with Almy and Brown. Slater warned his partners: "Do not converse with him of the subject before you hear the wole dispute."[26] Slater's authority within the factory was questioned. The line of authority and responsibility was clouded, and all three factions—Almy and Brown, Samuel Slater, and the workers—wanted to dominate and control factory operations.

Almy and Brown obviously distrusted Slater. Their mercantile training and experience account partially for their apprehension. In colonial America the general merchant had operated on a small scale, and usually within the framework of the family partnership. Most merchants, including the Browns of Providence, the Hancocks of Boston, and the Olivers of Baltimore, had chosen their partners from among kinsmen, close family friends, or long-standing business associates, and rarely from outside this closed network. In the partnerships formed, usually the proprietors not only owned the concern but also managed it, often working alongside their clerks and assistants in the office or the store. They believed that self-interest was the primary driving force in the successful operation of their business and that theft, deceit, and inattention to duty would result if one entrusted capital and goods to the care of outsiders. In these family firms management presented few difficulties, for merchants rarely supervised more than a handful of clerks, assistants, and supercargoes. Some of these employees were related to each other, and many of the others were drawn from the same social and economic classes occupied by their employers. It had been in this type of firm, in which ownership was synonymous with management and in which family connections usually meant more than skill in the employment of personnel and the choice of partners, that Almy and Brown had served their apprenticeships. They

26. Almy and Brown MSS, Samuel Slater to Almy and Brown, July 24, 1795, quoted in Kulik, "Beginnings of the Industrial Revolution in America," p. 211.

tried to transfer these lessons to the new enterprise.[27] They viewed Slater with disdain, and they tried to wrestle all authority from him. In fact, they were able to diminish his financial interest in Pawtucket when in 1797 they allowed him only a one-third share, not a full half share, of the Pawtucket lot and mill privileges.[28]

MARKETING

Other aspects of their business, especially the marketing of goods, proved less problematic for Almy and Brown than the production of yarn. They disposed of all yarn. Part of it was destined for their weaving workshop. The largest textile manufacturers in Providence, they employed twenty-three weavers in their workshop in 1791 and manufactured more than 12,000 yards of cloth, including fustians, velverets, janes, and denims.[29] Their workshop, however, seldom operated at full capacity: "we having three looms pretty much idle for want of a workman [and] our apprentices not being capable of carrying on the work," they complained to a friend in 1791.[30] Because the partners could not secure an adequate supply of qualified weavers and apprentices, yarn began to accumulate. In 1791 they considered abandoning their workshop and putting their apprentices into the spinning mill. In a letter to Barnabas Allen, father of an apprentice, they suggested that his son be transferred from the stocking to the spinning trade: "Stocking weaving has not as yet been adequate. . . . We think [it] not best to revive the stocking weaving we have thought of his beginning soon to Spin."[31] Almy and Brown kept the workshop open, but they failed to solve their labor shortage problems. Repeatedly Smith Brown traveled to Norwich and to other communities in search of workers "at the

27. W. T. Baxter, *The House of Hancock: Business in Boston, 1724–1775* (New York, 1965 [1945]), pp. 147, 199, 295, 304; Stuart Bruchey, *Robert Oliver, Merchant of Baltimore, 1783–1819* (Baltimore, 1956), pp. 58, 84, 166–167; Hedges, *Browns of Providence Plantations,* pp. 289–290, 328; Chandler, *Visible Hand,* pp. 9, 14; see also Pollard, *Genesis of Modern Management,* pp. 6–12.

28. Kulik, "Beginnings of the Industrial Revolution in America," p. 165.

29. Ibid., pp. 149–150; Conrad, "Evolution of Industrial Capitalism in Rhode Island," pp. 73–74.

30. Almy and Brown MSS, William Almy to Friend, March 4, 1791.

31. Ibid., Almy and Brown to Barnabas Allen, May 25, 1791; Allen to Almy and Brown, June 2, 1791.

stocking weaving business."[32] The workshop struggled on for several more years until around 1796 Almy and Brown closed it down. They then set up as putting-out agents, using their Providence store as their headquarters. Yarn was put out to local female hand-loom weavers in the area who "do not in general follow the occupation regularly; it is done during their leisure hours, and at the dull times of the year."[33] Yarn was also put out to workshop weavers in the vicinity. The pay for weaving men's stockings was between 3s. and 3s. 6d. per pair and for women's stockings between 2s. 3d. and 3s. per pair. The putting-out trade expanded until by 1809 approximately one hundred people wove for Almy and Brown. The partners finished the cloth in their dye house and calender building.[34]

Despite the growth of their putting-out business, Almy and Brown could not use all of the yarn produced by Samuel Slater, and they had to develop a market for their surplus goods. The organization and the network they established proved successful, and this success assured the continued growth and prosperity of their textile business. Certainly their mercantile background appeared to suit them well for this task. The Brown family had commercial interests throughout New England and the Middle Atlantic states which initially served the new firm well. Almy and Brown's connections with the wholesale and import business of Nicholas Brown and Company figured prominently in their plans. With outlets in Providence and Grafton, Nicholas Brown and Company sold goods to shopkeepers in Coventry, Cumberland, Smithfield, Uxbridge, Mansfield, Sutton, New Providence, Woodstock, Killingly, East Guilford, Willington, and Stonington.[35] It was during this era and through the wholesale and import arm of this business that the Brown family established personal and commercial connections with shopkeepers and wholesalers throughout the Northeast. Now Almy and Brown used these connections.

32. Ibid., Louis Bard to Brown, April 2, 1792; William Thompson to Almy and Brown, July 23, 1791.
33. Henry B. Fearon, *Sketches of America* (London, 1818), p. 100; Samuel Slater and Sons, Inc., *The Slater Mills at Webster* (Worcester, n.d.), p. 21.
34. Almy and Brown MSS, Almy and Brown to Thompson, August 9, 1791; Adam Duggan to Almy and Brown, May 15, 1793; see also Moses Brown, May 1, 1809, quoted in Clark, *History of Manufactures*, 1:432.
35. Hedges, *Browns of Providence Plantations*, pp. 12–14, 187.

With their family reputation for honest trading behind them, with their knowledge of American commercial and business practices to guide them, and with a battery of shopkeepers drawn from their various other commercial accounts to serve initially as their selling agents, Almy and Brown set about vending their machine-manufactured yarn.[36]

To acquaint potential customers with their product, they advertised in urban and country newspapers. One advertisement placed in a Portsmouth, New Hampshire, newspaper read:

> ALMY & BROWN, Providence, State of Rhode Island, being principally concerned in the most extensive COTTON MILLS on the Continent, have for sale a large assortment of Cotton Yarn suitable for Warp or Filling, and also of Two and Three Threaded Yarn, suitable for knitting or weaving Stockings, of equal quality to any in America; they are enable to supply their customers, and others who may want, on as low terms as any person in the United States, and on short notice.[37]

Furthermore, Obadiah Brown, who replaced Smith Brown as a partner in the firm, traveled up and down the coast soliciting customers. This aggressive marketing technique brought results. In 1797 Brown collected receipts for nearly $600 from J. Burnham of Salem and also assayed the possibilities of vending yarn in Marblehead.[38]

Among country storekeepers, factory-spun yarn became a popular item. Daniel Waldo, a Worcester shopkeeper, wrote that "the savin it will make in Private families . . . begins to be generally known in their neighborhood; the consequence is that the demand has increased very sensible."[39] Abner Greenleaf of Newburyport reported a similar experience: "As to the sale of the cotton, we have sold about one half of what you sent us, and it seems to be approved by our women, much more since they have made use of it

36. Ware, *Early New England Cotton Manufacture*, p. 32; Almy and Brown MSS, Almy and Brown to James Bringhurst, February 13, 1797.

37. Bagnall, *Textile Industries of the United States*, p. 163.

38. Almy and Brown MSS, Obadiah Brown to Almy and Brown, April 25, 1797; and Ware, *Early New England Cotton Manufacture*, p. 36.

39. Almy and Brown MSS, Daniel Waldo to Almy and Brown, April 27, 1801, quoted in Ware, *Early New England Cotton Manufacture*, p. 32.

than before they had tried it and we now sell it faster than at first."[40] Many of these shopkeepers later abandoned their retail sales and became merchant manufacturers, the central agents in the domestic production of cloth.[41]

Almy and Brown found that, in addition to shopkeepers, professional weavers wanted their yarn. In 1793 Peter Colt wrote: "I am desired to write you respecting cotton thread suitable for the stocking weaving business, for the purpose of supplying *Eight Looms* for some months at the Factory at Paterson in New Jersey. . . . In your proposals you will please describe the fineness of your yarns in such a manner that the Head workman can judge of it without seeing a sample, that we may know what number to order in case the terms suit."[42]

Although Peter Colt did not want to see samples of the yarn, most customers did. When a potential customer requested information, Almy and Brown sent samples of the various yarns manufactured and stipulated the terms on which they would agree to send bulk orders. Supplying each customer with the precise yarn required was a difficult and a time-consuming task, especially when, as Almy and Brown confessed to Dan Deshon in 1798, "our acquaintance with . . . looms, is not sufficient to enable us to assertain with precision what size of yarn will best suit them."[43] Several letters and packets of yarn passed between buyer and seller before arrangements were concluded. If the yarn answered the customers' requirements, they could purchase and sell it on their own account or take it on consignment. Most customers chose the latter proposal. Consignment sales became the rule. Commission rates varied from 2.5 to 5 percent, credit was extended from three to nine months, and Almy and Brown assumed all risk for unsold goods.[44]

40. Almy and Brown MSS, Abner Greenleaf to Almy and Brown, September 6, 1797.

41. Ibid., Thomas Hazard to Almy and Brown, January 17, 1798; Almy and Brown to Dan Deshon, October 13, 1798; Abner Greenleaf to Almy and Brown, September 6, 1797; Pollard, *Genesis of Modern Management*, pp. 11–12.

42. Almy and Brown MSS, Peter Colt to Almy and Brown, May 22, 1793.

43. Ibid., Almy and Brown to Deshon, October 13, 1798.

44. Ibid., Edmund Badgen to Almy and Brown, September 21, 1797; Almy and Brown to Colt, May 31, 1793; Almy and Brown to Deshon, October 13, 1798; Almy and Brown to Bringhurst, February 13, 1797; and Clark, *History of Manufactures*, 1:365–366.

Almy and Brown, however, did not have the American market to themselves. Both foreign and domestic manufacturers proved adept competitors. The United States imported vast quantities of British yarn and cloth. British exports of printed cloths and linen alone to the United States climbed steadily, from 353,762 yards in 1785 to 3,710,471 yards in 1800.[45] To retain their position in the American market and perhaps to eliminate their new American competition, British mills supplied customers with quality goods at low, competitive prices. Furthermore, to boost sales, they offered their local agents high commission rates and allowed them long-term credit. Almy and Brown had to match British terms and to lower prices if they wanted to attract customers. A Boston merchant house, Parkman and Black, told them in 1792 that "if you can make the price agreeable, we are always willing to promote the manufacturers of our own country."[46] Almy and Brown complied and lowered their prices, but they felt pressed by these merchants. "You may be assured that at the price they are marked there is but barely a living profit upon the making them as they are much heavier than English Jeanetts of the same superficial appearance," the Providence partners replied to the Boston firm.[47] To compound problems, Almy and Brown faced limited competition from American producers who had emerged that decade. They, too, tried to undersell the partners.[48] In cloth sales, for example, Almy and Brown competed with goods manufactured by workshop weavers and by individuals in their homes.[49]

Throughout the decade, the partners worked hard to attract and retain customers. Competitive prices, attractive terms, attention to the demands and requirements of customers, and quality goods

45. Elizabeth B. Schumpeter, *English Overseas Trade Statistics, 1697–1808* (London, 1960), p. 67.

46. Almy and Brown MSS, Parkman and Black to Almy and Brown, November 25, 1792, quoted in Conrad, "Evolution of Industrial Capitalism in Rhode Island," p. 79.

47. Almy and Brown MSS, Almy and Brown to Parkman and Black, April 9, 1793.

48. Ibid., William Dean to Almy and Brown, April 6, 1798. Dean wrote: "We think there is a prospect that we might have occasion to order largely of cotton wick yarn if you can sent it @ 3/6 . . . but not at any higher price as they manufacture it at Beverly for 3/ however as yours is better we are in hopes we might obtain the preference at 6 difference." See also Hazard to Almy and Brown, January 17, 1798.

49. Ibid., Colt to Almy and Brown, May 22, 1793; Almy and Brown to Thompson, Providence, August 9, 1791; Dugan to Almy and Brown, May 15, 1793.

allowed them to expand their share of the market. Toward the end of the decade they sold goods in communities large and small from Portland to Baltimore. Their list of shopkeepers, workshop owners, merchants, and manufacturers grew until they no longer could serve adequately the demands of customers. At this junction, they tried to consign their products to a limited number of retailers and wholesalers in the major port cities. In letters to kinsmen, business associates, and fellow Quakers, they requested information on wholesalers and retailers.

To James Bringhurst, a Philadelphia Quaker, they wrote in February 1797: "We were wishing to introduce the sale of cotton yarn, in your City and knowing of any person who would be suitable, and would probably undertake to sell it for us we have ventured to take the liberty to ask thy favor of thy nameing some friends who was in the business of retailing dry goods and who would be suitable and who would be willing to make tryal."[50] They were put in touch with Elijah Waring, a local domestic wholesaler who subsequently became their leading vendor in Philadelphia. Not only did Almy and Brown require Waring to supply his traditional retail customers with yarn, but they also requested that he solicit professional weavers and other customers. In 1797 Waring took samples of yarn to Germantown, where he "met with very great discouragements, the weavers there being all poor men and principally engaged in worsted."[51] He counseled the Providence partners that yarn for darning and knitting would be suited better to the Philadelphia market and that shopkeepers, not small workshop proprietors, would be the most appropriate customers.[52] In the larger urban centers, transactions passed increasingly through wholesalers. These men ascertained the ply and the color of yarn suitable to the needs of the customers they solicited and advised the Providence partners on fashion trends, prices, and competition in their market.[53]

By 1799 Almy and Brown felt confident enough in their product

50. Ibid., Almy and Brown to Bringhurst, February 13, 1797.
51. Ibid., Elijah Waring to Almy and Brown, May 22, 1797.
52. Ibid., Almy and Brown to Waring, November 3, 1797.
53. Ibid; see also Glenn Porter and Harold C. Livesay, *Merchants and Manufacturers: Studies in the Changing Structure of Nineteenth-Century Marketing* (Baltimore, 1971), pp. 24–25; Ware, *Early New England Cotton Manufacture*, pp. 167–169.

and in its general acceptance by the public to introduce two price rises: a 5-cent-per-pound increase in May and a 10-cent-per-pound increase in October.[54] Demand continued to rise. By 1800 Almy and Brown notified customers that requests have "increased so that it is not at present in our power to supply it."[55] Retailers sold out quickly, and they "could have sold a great quantity" more if they had had it on hand.[56]

With demand high, Almy and Brown decided to expand their holdings. In 1799 they bought an interest in the Warwick Spinning Company and sent one of their new mechanics to Pawtucket to take measurements of Slater's equipment.[57] They did not invite Slater into the new firm. Now that they had access to technical knowledge, they could do what they had wanted to do a decade earlier— operate on their own without the assistance of an outsider.

In financial terms the Almy, Brown, and Slater partnership proved successful. Together these entrepreneurs demonstrated that machine-produced yarn could be manufactured in the United States and that there was a ready market for that yarn and for the cloth and stockings made from it. The internal management of the business, however, proved problematic. The British form exemplified by the Strutt, Need, and Arkwright partnership failed to work well initially in the United States. Accustomed to dealing primarily within kinship and religious networks, American merchants such as Almy and Brown could not entrust funds or responsibility over factory operations to an outsider. For almost a decade, Almy and Brown had treated Slater with indifference and even contempt. Under their domination, Slater had not been able to develop fully the type of factory system he knew would operate effectively in the United States. With authority over yarn production divided between two antagonistic forces, smooth, efficient operation of the mill was practically impossible. All decisions had to be brought under the control and

54. Conrad, "Evolution of Industrial Capitalism in Rhode Island," p. 135.
55. Almy and Brown MSS, Almy and Brown to Smith and Allerson, June 27, 1800, quoted in Conrad, "Evolution of Industrial Capitalism in Rhode Island," p. 134.
56. Almy and Brown MSS, William Stevens to Almy and Brown, November 10, 1800, quoted in Conrad, "Evolution of Industrial Capitalism in Rhode Island," p. 134.
57. Bagnall, *Textile Industries of the United States*, p. 217.

direction of a single individual or group of like-minded people. While Slater struggled against a force determined to dominate all phases of factory production, he faced a second challenge to his authority from another direction: his labor force tried to gain control of the factory floor.

[3]

Samuel Slater and the Pawtucket Factory

Economic changes do not lead necessarily to the immediate destruction of traditional values. Recent research on the stage of development preceding modern industrialization, which some historians have labeled proto-industrialization, has confirmed that traditional values associated with a rural, subsistence way of life persisted, and in some cases flourished, within the context of a growing market economy. The family especially proved resilient during periods of economic upheaval. In a persuasive series of essays on the proto-industrial family economy, Hans Medick argued that during this phase, "production, consumption, and generative reproduction increasingly broke away from their agrarian base. They came to be entirely determined by the market, but, at the same time, they preserved the structural and functional connection that was provided by the family." He concluded that "the social mode of the 'ganzes Haus' still formed an effective socio-economic structural model, after its agrarian subsistence base had largely disappeared."[1] Medick's work emphasized pressures placed on the European family during the initial movement from a peasant to a household economy.

American historians also have been drawn to the transitional era between a subsistence way of life and the rise of the factory system. Toward the middle of the eighteenth century in the northern

1. Hans Medick, "The Proto-Industrial Family Economy," in *Industrialization before Industrialization*, ed. Peter Kriedte, Hans Medick, and Jurgen Schlumbohm (Cambridge, Eng., 1981), p. 40.

American colonies, a commercially oriented agricultural economy began to emerge, and household and workshop production of cloth, hats, buttons, boots, shoes, and other goods became significant. People participated increasingly in a growing market-oriented economy.

James Henretta examined the changing economy and its effect on the world view of the participants, especially the family. He concluded:

> Even as this process of economic specialization and structural change was taking place, the family persisted as the basic unit of agricultural production, capital formation, and property transmission. This is a point of some importance, for it suggests that alterations in the macro-structure of a society or an economic system do not inevitably or immediately induce significant changes in its micro-units. Social or cultural change is not always systemic in nature, and it proceeds in fits and starts. Old cultural forms persist (and sometimes flourish) within new economic structures; there are "lags" as changes in one sphere of life are gradually reconciled with established values and patterns of behavior.[2]

Those who lived during the economic transformation of this period "continued to view the world through the prism of family values."[3]

The family and traditional values withstood the test of another economic transformation, the rise of the factory system. The factory system established by Samuel Slater did not necessarily destroy old customs. Within the context of a new economic order, the values associated with New England family life could be observed. The transition between a preindustrial way of life and modern industrialization was a drawn-out process.

LABOR AND THE NEW ENGLAND ECONOMY

When Samuel Slater assumed his position as primary technician and partner in the Pawtucket factory, the New England economy

---

2. James Henretta, "Families and Farms: *Mentalité* in Pre-Industrial America," *William and Mary Quarterly* 35 (January 1978):25.
3. Ibid., p. 32.

was in transition. In the last half of the eighteenth century, the agricultural sector had reached its natural limits. Inflation, population growth, a decline in the size of farm holdings, poor agricultural techniques, and the economic dislocation caused by several wars had adversely affected rural agricultural towns. This was especially true in eastern New England, where population pressure together with declining landholdings limited economic opportunities. Parents found it difficult to provide an adequate settlement for their children. Many could give them neither land nor money; nor could they guarantee them an education that would prepare them for a profession or a trade. Their children faced two alternatives: they could enter the temporary, casual labor market or they could migrate. Out-migration to the less populous states of Maine and New Hampshire or to the western regions of New England offered one route of escape that many chose to take. While out-migration might mean economic advancement, those who left often paid a high price. Leaving family and friends, leaving the community in which they had grown to maturity, moving to another region, to a strange town where they knew few people, could cause considerable stress. Migration also could be expensive: land had to be purchased, and seed, tools, animals, and household furnishings had to be purchased or transported. Furthermore, there were no guarantees that the move would be economically advantageous. The rural economy of newly settled areas might prove disappointing, and the move might be merely the first of many.[5]

Entry into the urban labor market offered another alternative. Yet the economic opportunities available in the towns were often elusive; unskilled laborers were not in demand. In commercial towns large and small, such as Beverly, Salem, and Boston, an increasingly stratified society with discernible class divisions took shape. In 1741 the wealthiest 10 percent of the population of Beverly, at that time a growing commercial center, owned 39 percent of the real estate and about 45 percent of the personal property, includ-

4. Greven, *Four Generations*, pp. 222–258; Faler, "Workingmen, Mechanics, and Social Change," pp. 6–23; Jones, "Geographic Mobility and Society in Eighteen-Century Essex County," pp. 161–200.

5. Greven, *Four Generations*, pp. 155–172; see also Jones, "Geographic Mobility and Society in Eighteenth-Century Essex County," pp. 176, 161, 205–206.

ing agricultural implements, ships, businesses, and shops. Following the Revolutionary War, the gap between the rich and the poor widened; the amount of real property owned by this group rose to 49 percent, while their hold on personal property reached 79 percent. By then approximately 40 percent of the people in Beverly owned little or nothing. Those who migrated to this town found few reasons to remain, and many moved on to other communities. The situation was no better in the larger commercial port cities, where wealth was even more concentrated.[6] By the time of the Revolutionary War a large migratory population was constantly changing places at the lower levels of society.[7]

At the war's end, with the onset of a general depression, residents of rural interior communities faced a potentially disastrous future. Local economies were in shambles. Many communities faced a tax increase to pay for their share of the war effort and to support the rising indigent population. In one community after another the poor became a major problem, and stiff measures were undertaken to control them. The war years and their aftermath had depleted the finances of Oxford, Massachusetts, while the number of indigent and disabled people continued to grow. The town's historian, George Daniels, believed that "the number of indigent persons had largely increased in consequence of the Revolutionary war, and public burdens were oppressive."[8] The situation became more acute as increasing numbers of dispossessed and displaced people arrived after the war to look for work. Town officials reacted to this growing population of indigents by warning out those they considered disruptive or a possible future charge on the town. In 1789 a Sutton family and an ummarried woman from Ware, Massachusetts, were ordered to leave, signaling the begin-

6. Jones, "Geographic Mobility and Society," pp. 161–199; Donald W. Koch, "Income Distribution and Political Structure in Seventeenth-Century Salem, Massachusetts," *Essex Institute Historical Collection* 105 (January 1969):61, 63; Allan Kulikoff, "The Progress of Inequality in Revolutionary Boston," *William and Mary Quarterly* 28 (July 1971):376. For another interpretation see Gloria L. Main, "Inequality in Early America: The Evidence from Probate Records of Massachusetts and Maryland," *Journal of Interdisciplinary History* 7 (Spring 1977):560–568, 572–573.

7. It is arguable that such migratory populations also reflected harvest demands, that workers moved to towns when they were not needed on the farm. I am grateful to Paul G. E. Clemens for this observation.

8. George F. Daniels, *History of the Town of Oxford, Massachusetts, with Genealogies and Notes on Persons and Estates* (Oxford, Mass., 1892), pp. 769–770.

ning of a purge of the poor and unwanted population. In February 1792 seventy-eight people were warned out, and in June another twenty-three were told to leave; forty-two persons were ordered out of town in 1793.[9] Some of the larger port cities also adopted the practice of wholesale warnings. In June 1791 the Salem town selectmen told 261 householders to leave with their families and dependents within fifteen days. Throughout the commonwealth the practice was repeated; anyone who returned faced a fine, corporal punishment, incarceration in an almshouse or workhouse, or worse. In some communities maintenance of the poor was provided by vendue, "they being put out for the year to the lowest bidder."[10] Their employer provided them with clothing, food, lodging, medical care, and "all such necessaries fitting for them in their degree!"[11] At least one-third, and possibly more, of the transient poor were liable to such treatment. Elsewhere many poor children were sent to work as apprentices.

The pauper apprenticeship scheme was designed specifically for indigent, illegitimate, and orphaned children. Taken from their parents or guardians, these children were apprenticed to "orderly families." About half of the children subject to training were under the age of nine. Like formal trade apprenticeship programs designed to train and educate skilled workers, the employment of these children was regulated by legislation. Each child served a specific length of time, had to learn a trade or occupation, and had to be taught to read and to write and, in the case of males, to cipher.

9. Ibid.

10. Thomas Steere, *History of the Town of Smithfield from Its Organization, in 1730–1, to Its Division in 1871* (Providence, 1881), p. 51. Many communities gathered the poor together and housed them in an almshouse or workhouse, located usually at a considerable distance from the community center. Some smaller villages even set aside building for the incarceration of the transient poor. In Douglas, Massachusetts, for example, the old schoolhouse was refurbished and turned into a workhouse "where people were confined and put to hard labor for the crime of poverty." The town's historian noted: "Its use for such a purpose made it an object of special aversion to all whom necessity had made the subjects of public charity" (see William A. Emerson, *History of the Town of Douglas, Massachusetts, Earliest Period to the Close of 1878* [Boston, 1879], p. 55). See also "Salem Warnings, 1791," *Historical Collections of the Essex Institute* 43 (October 1907):345–352; Josiah Henry Benton, *Warning Out in New England* (Boston, 1911), pp. 46–62; Daniels, *History of the Town of Oxford*, pp. 222–223; Robert W. Kelso, *The History of Public Poor Relief in Massachusetts, 1620–1920* (Montclair, N.J., 1969), pp. 50–51.

11. Steere, *History of the Town of Smithfield*, p. 51.

Unlike trade apprenticeship, however, this scheme was not voluntary. Boston's pauper apprenticeship system, in which boys entered such occupations as baking, smithing, seafaring, and farming, demonstrates the diversity possible within the program. While before the 1760s the crafts absorbed most of the young male apprentices, thereafter husbandry became the primary occupation. In the 1770s approximately 60 percent of the boys were bound out to farmers, and in the ensuing two decades, 55 percent of such boys entered husbandry. Indigent girls also faced apprenticeship training, but their area of employment was chiefly housewifery.[12] In rural communities occupations for pauper apprentices would naturally be restricted largely to farming and household duties. In fact, it appeared that rural residents could not secure enough of these hands to satisfy local demands, and Boston came to serve as a labor reservoir for much of the region. In the 1790s approximately 82 percent of Boston's pauper apprentices were bound over to families in Suffolk, Middlesex, Hampshire, and Worcester counties.[13]

In the late eighteenth century many people lived at the economic margins of society. Approximately 10 percent of the population of New England were referred to as transients, vagabonds, and "unwanted persons"; these people were on the tramp in the region, moving from town to town in search of work and a permanent settlement. They possessed little or no real estate or personal property, were destined to remain poor, and were subject to increasingly harsh poor laws. It has been estimated that in the larger cities, such as Boston, another 37 to 47 percent of the population were classified as the "near poor": many widows, blacks, unskilled laborers, sailors, and agricultural workers straddled the line between subsistence and destitution.[14] It was from among these poor people that Samuel Slater recruited workers.

12. Lawrence W. Towner, "The Indentures of Boston's Poor Apprentices: 1734–1805," Publications of the Colonial Society of Massachusetts, vol. 43, *Transactions, 1956–1963* (1966), pp. 425–426, 435–449; Douglas Lamar Jones, "Poverty and Vagabondage: The Process of Survival in Eighteenth-Century Massachusetts," *New England Historical and Genealogical Register* 133 (October 1979):250.

13. Kelso, *History of Public Poor Relief in Massachusetts*, pp. 91–97; Towner, "Indentures of Boston's Poor Apprentices," pp. 417–433.

14. Kulikoff, "Progress of Inequality in Revolutionary Boston," pp. 383–386; Jackson Turner Main, *The Social Structure of Revolutionary America* (Princeton, 1965), pp. 22–23; Charles Grant, *Democracy in the Connecticut Frontier Town of Kent* (New York, 1961), p. 97.

For Slater construction of machinery in the Arkwright style presented few problems. Assisted by several local craftsmen, he built three carding machines, one carding and roving machine, and two water frames, one of twenty-four spindles and another of forty-eight spindles. He temporarily installed this equipment in a clothier's shop in Pawtucket, Rhode Island, and hired four boys to operate it. Within three weeks he doubled the number of child operatives, and girls as well as boys found work there. These children, some as young as seven years old, performed a wide variety of simple tasks, from picking the cotton clean of dirt, leaves, pods, and other foreign matter to operating the carding, roving, and spinning machines.[15] One of the first factory operatives, Smith Wilkinson, later described his duties:

> I was then in my tenth year, and went to work with him, and began attending the breaker. The mode of laying the cotton was by hand, taking up a handful, and pulling it apart with both hands, shifting it all into the right hand, to get the staple of the cotton straight, and fix the handful, so as to hold it firm, and then applying it to the surface of the breaker, moving the hand horizontally across the card to and fro, until the cotton was fully prepared.[16]

These young children were only a temporary labor force and were primarily the sons and the daughters of local artisans in the area. Smith Wilkinson, for example, was the son of Oziel Wilkinson, a local iron manufacturer. Slater lived with the Wilkinson family and eventually married Hannah Wilkinson, a daughter in the house-

15. Frederick L. Lewton, "Samuel Slater and the Oldest Cotton Machinery in America," *Annual Report of the Board of Regents of the Smithsonian Institution, Showing the Operations, Expenditures, and Condition of the Institution for the Year Ending June 20, 1926* (Washington, 1927), p. 509; Jonathan T. Lincoln, "Beginnings of the Machine Age in New England: David Wilkinson of Pawtucket," *New England Quarterly* 6 (December 1933):721–722; John A. La Porte, "Birth of America's Spinning Industry, II," *New England Magazine* 29 (February 1909):676; E. H. Cameron, "The Genius of Samuel Slater," *Technology Review* 57 (May 1955):336–337; White, *Memoir of Samuel Slater*, p. 99; Ware, *Early New England Cotton Manufacture*, pp. 22–23; Browne, *Genealogy of the Jenks Family*, pp. 109, 211; Almy and Brown MSS, Samuel Slater to Almy and Brown, March 13, 1792.

16. *Report of the Centennial Celebration of the Twenty-Fourth of June 1865, at Pawtucket, of the Incorporation of the Town of North Providence* (Providence, 1865), pp. 32–33.

hold.[17] A second operative, Jabez Jenks, aged ten, was the son of a carpenter who helped construct buildings and a dam for Almy, Brown, and Slater.[18]

When construction of a permanent building and its ancillary facilities began, Slater slowed production in the clothier's shop and devoted most of his energies to the new task. He either supervised directly or retained a keen interest in the construction of a wood-frame dam across the Blackstone River which was finished in 1792, a purpose-built factory, and perhaps some additional equipment. Completed in 1793, this two-story mill with a large attic measured 43' by 29'. The carding and the spinning departments were separated, and windows encircled the rooms. According to Gary Kulik, an authority on early mill construction, "the small, 'factory house' did not look noticeably different from the buildings around it. Wholly traditional in its use of post and beam construction, the mill resembled both domestic and institutional forms. If it indeed had an end belfry, the mill would have appeared similar to an unpretentious New England meeting house—a demurely sober and comforting image."[19]

Construction of the mill and the dam required several years, and in 1793 Slater was ready to begin operations in earnest. When work on the mill project was completed, Slater dismissed many of the construction workers, and because the sale of yarn was slow, he laid off many of the children and reduced the wages of others.[20] As a result, many householders packed up and left, seeking work in neighboring communities. When Slater wanted to start up his factory, many of the householders who had built the mill and whose children had operated the first machines refused to return to Pawtucket unless Slater guaranteed them steady work. But Slater did not need skilled workers. Retaining their parents in some capacity

17. Gilbane, "Social History of Samuel Slater's Pawtucket," pp. 75–93; Browne, *Genealogy of the Jenks Family*, pp. 109, 211.

18. Robert Grieve and John P. Fernald, *The Cotton Centennial, 1790–1890: Cotton and Its Uses, the Inception and Development of the Cotton Industries of America, and a Full Account of the Pawtucket Cotton Centenary Celebration* (Providence, 1891), p. 28; Browne, *Genealogy of the Jenks Family*, p. 109.

19. Kulik, "Beginnings of the Industrial Revolution in America," p. 160; See also Bagnall, *Textile Industries of the United States*, pp. 159–160; La Porte, "Birth of America's Spinning Industry, II," p. 677.

20. Almy and Brown MSS, Account of Labour: Children's Work, 1792.

in or around the mill was expensive, but unskilled, inexperienced child laborers could easily operate the simple machines. Almy, Brown, and Slater turned to child labor.[21]

Children seemed an ideal choice. Wages paid to children were low; most received from 2s. to 5s. weekly, depending on their job. Compared to the prevailing rate of approximately 3s. per day for adult male agricultural labor, this was a bargain.[22] Furthermore, contemporary social attitudes encouraged the employment of youngsters. Alexander Hamilton's often quoted remark is typical of the prevailing attitude in the post-Revolutionary era. He argued that factory masters should employ

persons who would otherwise be idle (and in many cases a burden on the community), either from the bias of temper, habit, infirmity of body, or some other cause, indisposing or disqualifying them for the toils of the country. It is worthy of particular remark, that, in general, women and children are rendered more useful, and the latter more early useful, by manufacturing establishments, than they would otherwise be. Of the number of persons employed in the cotton manufactories of Great Britain, it is computed that four sevenths, nearly, are women and children; of whom the greatest proportion are children, and many of them of a tender age.[23]

His sentiments were echoed by countless other respected members of the community. A writer for the *American Museum* encouraged parents to teach children a skill or to apprentice them out to a trade, for unless they were constantly at work, they would stray and become candidates for the gallows.[24] Enos Hitchcock, an eighteenth-century author of child-guidance books, concurred: "In order to prepare children to live in the world, it is necessary to train them up in the knowledge of business, and in the habits of industry: . . . Childhood and youth may be considered as a term of ap-

21. Conrad, "Evolution of Industrial Capitalism in Rhode Island," p. 99.
22. Almy and Brown MSS, Account of Labour: Children's Work, 1792; see also Main, *Social Structure of Revolutionary America*, p. 70.
23. Alexander Hamilton, "Report on Manufactures," December 5, 1791, *Works* (New York, 1850), cited in *Children and Youth in America*, 3 vols., ed. Robert H. Bremner (Cambridge, Mass., 1970), 1:172.
24. "The Child Trained Up for the Gallows," *American Museum*, February 1791.

prenticeship in which they are to be exercised in those employments whereby they are to live."[25]

In trying to find a suitable form of labor, Slater sought guidance from the past. Apprenticeship labor and child wage workers both presented viable solutions. These schemes had operated effectively in England, and Slater saw no reason for them not to take root in Rhode Island, as well. From 1794 to the end of the decade pauper apprentices and child wage workers labored together in his factory. In the end, however, neither form of labor proved appropriate.

APPRENTICESHIP LABOR

An advertisement published in the *Providence Gazette* signaled Slater's decision to employ apprentices to operate some of his equipment; he wanted "four or five active Lads, about 15 Years of Age to serve as Apprentices in the Cotton Factory."[26] In response to this notice, letters arrived from both private citizens and overseers of the poor offering to bind out their young male charges. As revealed through the firm's correspondence, poverty motivated people to apprentice children. "Observing your advertisements in the papers for Boys to work in your manufactory," wrote one Rhode Island woman, "I take the liberty of sending one boy the name of Thomas Tippets, who I wish may answer your purposes as he is at present destitute of a home and very poor."[27] In another case, the selectmen of Rehoboth, Massachusetts, offered to bind one of their young, indigent charges, James Horton, to Slater. The boy said that "he (himself) was willing."[28]

The factory became an employer of pauper apprentices, and this scheme initially appeared to be a viable solution to Slater's labor problem. The apprenticeship system was in place in scores of New England towns and was accepted by the community as a mechanism for the education and support of youngsters. By appealing to

25. Enos Hitchcock, *The Farmer's Friend, or the History of Mr. Charles Worthy* (Boston, 1793), pp. 131–132.
26. *Providence Gazette*, October 11, 1794, quoted in Gilbane, "Social History of Samuel Slater's Pawtucket," p. 247.
27. Almy and Brown MSS, Sally Brown to Sarah Brown, November 14, 1794.
28. Ibid., Samuel Slater to Almy and Brown, May 25, 1797.

poor-law officials and indigent householders, manufacturers could not only choose operatives from a large pool of potential workers but also provide a public service by lowering the cost of maintaining paupers.

While in theory the apprenticeship system had considerable merit, it proved a disaster at Pawtucket. Slater found the system inflexible, expensive, and inefficient. Like other employers who recruited pauper apprentices, Slater had to conform to certain guidelines governing the training and education of his apprentices. Right from the start, however, he found that he could not fully comply with the guidelines. He was not the traditional master craftsman who maintained close contact with his apprentices both inside and outside the factory and who transferred his skill and knowledge to them. Quite the contrary, throughout his tenure with Almy and Brown he endeavored to prevent others from acquiring his knowledge of machine construction and factory operations. As Josiah Quincy, a visitor to the Pawtucket mill in 1801, observed: "We found the proprietor very cautious of admitting strangers to view its operations, nor would he grant us the privilege until he had received satisfactory assurances that we were as ignorant and unconcerned about every thing relating to the cotton manufacture as he could wish."[29] Other commitments to the children and their guardians proved equally difficult to fulfill.

Educational services had to be provided. Slater opened a Sunday school in which students from Brown University taught the factory children basic reading and writing skills.[30] Only two Testaments and three spellers survive to indicate the content of the lessons taught in the school, but because it was patterned after Strutt's Sunday school in Derbyshire, Slater's school probably emphasized the teaching of certain moral values, including obedience, honesty, temperance, punctuality, and deference. George White, Samuel Slater's friend and biographer, believed that lessons taught in this school "successfully combated the natural tendency of accumulating vice, ignorance and poverty. Such remedies not only prevented

29. "Account of Journey of Josiah Quincy," *Proceedings of the Massachusetts Historical Society* 4, 2d ser., 1887–1889 (Boston, 1889), p. 124.
30. Almy and Brown MSS, Almy and Brown to Benjamin Allen, December 12, 1796; see also Gilbane, "Social History of Samuel Slater's Pawtucket," pp. 304–306.

their occurrence but had a tendency to remove them, when they actually existed."[31] This sentiment was echoed by a local inhabitant of Pawtucket, David Benedict, who believed that "by the means of his school . . . morals of these children of misfortune were improved."[32]

Despite this acclaim, the school operated discontinuously during the early years of the factory system, and it did little either to inculcate moral values or to teach pauper apprentices reading or technical skills. Expenses for a classroom, furnishings, textbooks, supplies, heating, and the teacher's salary appeared an unnecessary burden on the firm's scant resources and a commitment Almy and Brown were not ready to make.[33]

Other expenses for the children mounted. In 1794 Slater had resorted to pauper apprentices because they were available and could be introduced into the factory with a minimum of fuss. While costs had not figured prominently in the planning stages of this program, they became a major consideration once it was set in place. The cost involved in the support of an apprentice in 1796 can be estimated. At that time room and board with a family in the area ranged between $1.44 and $2.00 weekly for each apprentice. Expenses for medical bills, clothing, shoes, and other items could add another $0.32 or more to the weekly bill. Almy and Brown paid £1 1s. 2d., or approximately $3.40, to cover the medical bills of one of their young charges, Caleb Spencer. This sum represented more than the weekly wage of an unskilled male householder and was a considerable drain on resources. At Slater's factory, unskilled men received approximately $2.75 weekly.[34] Conceivably the firm could

31. White, *Memoir of Samuel Slater*, p. 117; see also ibid., pp. 107–108.
32. David Benedict, *Fifty Years among the Baptists* (Boston, 1860), p. 312, quoted in Gilbane, "Social History of Samuel Slater's Pawtucket," p. 310; even Brown University's President Maxcy argued that the Sunday school presented a unique opportunity to introduce moral values to factory children. To persuade a university student to accept a teaching position at the Pawtucket school, he reasoned "with him on the opportunity he would have to do good in Pawtucket; stating that there had never been a school of any description there, and no place of worship, and probably no religious or moral instruction, certainly not of a public nature" (White, *Memoir of Samuel Slater*, p. 282).
33. Gilbane, "Social History of Samuel Slater's Pawtucket," pp. 304–310.
34. Almy and Brown MSS, Samuel Slater to Almy and Brown, Pawtucket, "For Boarding Apprentices," Caleb Spencer, Joseph Tabor, and Stephen Hopkins, 1795–1796; see also Conrad, "Evolution of Industrial Capitalism in Rhode Island," pp. 110–111.

spend between $1.75 and $2.32 a week, or between $90.00 and $120.00 annually, to support apprentices.[35]

Compared to the wages paid child contract workers, costs involved in the care and maintenance of apprentices were exorbitant. In 1796 child wage earners who lived with their families earned from 2s. 6d. to 4s. 3d. ($0.40 to $0.68) weekly. (See Table 1.) And wages were paid to contract workers only for time actually spent in the factory. In 1796 the factory was closed often or put on a short schedule—once in February, when Slater ran out of cotton, and again in April and October, when he was shorthanded. The hands sometimes caused the factory to close; in July, for example, workers left the mill to pick whortleberries.[36]

High costs involved in the support of the apprentices and Slater's inability to fulfill his agreement were only two of the several reasons that precluded the expansion of this system. The "active Lads" objected to their treatment. Appalled by the demanding work schedule that kept them employed for twelve to sixteen hours each day, six days a week; betrayed by the master craftsman who failed to instruct them in the "mysteries" of the trade; bored by the educational and religious training provided by the Sunday school, many absconded. As itinerant farm hands, teamsters, or casual day laborers they could support themselves and control their own time.

Table 1. Weekly earnings of members of Roger Alexander family, 1794–1796

	1794	1795	1796
Lusina	2s. 3d.	3s. 9d.	4s. 3d.
Lydia	2s. 2d.	3s. 6d.	4s.
Rubin	1s. 9d.	3s. 3d.	3s. 9d.
Nomia			2s. 6d.

Note: The Roger Alexander family is representative. At this period, 1s. equaled about $0.16.
Source: Almy and Brown MSS, Account to Roger Alexander, November 1794–April 2, 1796.

35. Almy and Brown MSS, Samuel Slater to Almy and Brown, Pawtucket, "For Boarding Apprentices," Caleb Spencer, 1795–1796; Caleb Spencer to Almy and Brown, June 20, 1796; Providence, March 17, 1796; see also Slater MSS, Spinning Mills to Almy and Brown, Pawtucket, vol. 1, February 12, 1798; January 16, 1800.
36. Almy and Brown MSS, Slater to Almy and Brown, Pawtucket, January 26, February 19, April 12, October 4, and October 17, 1796. See also Conrad, "Evolution of Industrial Capitalism in Rhode Island," p. 105.

Slater had not fulfilled his side of the agreement, so why should they? In March 1797 Slater warned his partners that James Horton, apprenticed by the Rehoboth selectmen, ran away, and another boy followed, and "again If it is suffered to pass, another will go tomorrow—& so on until they are all gone."[37]

While in Britain pauper apprenticeship solved part of the labor problem, in America apprentices were a hostile, expensive, uncooperative, and singularly inappropriate form of labor; Slater was content to end the experiment. Workers also applauded its demise. In a letter to Almy and Brown, James Horton, the runaway apprentice, explained why he left: "If Mr. Slater had tought me to work . . . all the different branches I should have ben with you now but instead . . . he ceepe me always at one thing and I might have stade there until this time and never new nothing. I don't think it probable that I shall return . . . for the business I follow now I think is much more bennefisheal to my Interest."[38] There were few compelling reasons for young single men to remain within the factory system, where they were taught few skills, could expect little advancement, and were forced to adhere to the strict rhythm and demands of machinery. Almost any situation represented an improvement.[39]

FAMILY LABOR

Turning away from orphans and abandoned or otherwise neglected boys, Slater recruited children on contract. Child wage

37. Almy and Brown MSS, Samuel Slater to Almy and Brown, March 1797, quoted in Gilbane, "Social History of Samuel Slater's Pawtucket," p. 263; Jones, "Poverty and Vagabondage," p. 248.

38. Almy and Brown MSS, James Horton to Almy and Brown, November 25, 1799, quoted in Kulik, "Beginnings of the Industrial Revolution in America," p. 236.

39. I have not included indentured servants, a form of bound labor employed throughout the colonial era, in this discussion, but Slater did pay the passage of several laborers in the 1820s. In 1829, for example, he paid Captain Jones $45 for transporting Sarah Platt and her child from Liverpool to Boston, but details of her obligation could not be found. Indeed, it may be assumed that this was not a formal indenture, but rather that Slater was acting on behalf of an employee or a friend. See Slater MSS, Samuel Slater and Sons, vol. 200, G. Jones to Samuel Slater and Sons, Boston, April 18, 1829.

workers comprised the bulk of Slater's work force when he first started producing yarn in the clothier's shop, even when he introduced apprentices. For years boys and girls under contract shared the workplace with apprentices and labored under similar conditions.

For twelve hours a day in winter, and fourteen to sixteen hours a day in summer, six days each week, boys and girls from seven to approximately thirteen years of age worked in the factories. On Sunday they marched to the Sabbath school alongside apprentices. Yet the work schedule of these children was not so rigorous as this timetable would indicate. At the Pawtucket mill, in one accounting period of approximately twelve weeks, Sarah, Stephen, Deborah, Ann, Amy, and Nathan Benchley worked about seventeen days, fifty-six days, fifty-nine days, fifty-one days, thirteen days, and fifty-eight days, respectively. Arnold Benchley, their father, worked one day.[40] Labor shortages, insufficient supplies, inadequate water power, slow demand for yarn, and weather conditions occasioned delays. On such occasions Slater sent contract workers home; while apprentices probably could have been set to work picking cotton or doing other simple chores, they too worked a shorter schedule.[41]

Slater's wage workers and apprentices shared other characteristics. Like their bound workmates, these contract children were indigent. But unlike James Horton and the other apprentices, they belonged to kinship units, and their parents, not guardians or poor-law officials, placed them in the factories. Prompted by necessity, parents seized the economic opportunity offered by the factory system; they hired out their children and used their children's wages to support the family and keep the family unit together.

In Britain poverty had pushed orphans and workhouse children into the factories; in America, poverty pushed entire families into the new economic order emerging at Pawtucket. Slater learned quickly that this class and this social unit would comprise a significant proportion of the labor force. Later his son Horatio N. Slater recounted a story concerning one of the early factory families:

40. Almy and Brown MSS, Account of Labour: Children's Work, 1792.
41. Ibid., Slater to Almy and Brown, March 13, April 17, and October 30, 1795; April 12, October 4, 1796.

Mr. Slater was obliged to seek families, and induce them to emigrate to Pawtucket. He found one Arnold, with a family of ten or eleven, living near the village of ———, in a rude cabin chiefly made of slabs, and with a chimney of stone. The roof of this comfortless structure sloped nearly to the ground, but it was the home of these hardy people. Mrs Arnold appreciated it fully, for when her husband consulted her on the proposed change, she insisted that Mr. Slater should give them as good a house as their old one.[42]

This was not an isolated incident. In 1797 Obadiah Brown notified the partners that families in Marblehead, Massachusetts, were candidates for factory employment: "This is a place very disagreably situated, being very rockey and the inhabitants appear to be poor their houses very much on the decline. . . . Children appearing very plenty."[43] Toward the end of the decade indigent families largely had replaced apprentices and had become the primary source of labor.

Two groups that sought factory employment for their children were widows and unskilled male householders. For many of these people industrial labor was an appealing alternative to community poor relief. Widows, who had few occupations open to them and who feared the consequences of accepting poor relief, turned to the factory master for work.[44] Throughout the seventeenth and eighteenth centuries, housewifery served as the chief occupation for women. While during married life they might have assisted husbands in the management of the farm or in the domestic manufacture of such goods as shoes and hats, they performed these duties at home. The home served as an effective boundary within which women exercised considerable responsibility but beyond which they seldom ventured. Whether widowed or never married, a woman who wanted to work outside the home found that employment was limited to a few poorly paid, low-status occupations, such as midwifery, domestic service, and washing. These occupations could not

42. H. N. Slater's Reminiscences of Samuel Slater, April 26, 1884, quoted in William B. Weeden, *Economic and Social History of New England: 1620–1789*, 2 vols. (Boston, 1890), 2:913.

43. Almy and Brown MSS, Obadiah Brown to Almy and Brown, April 25, 1797.

44. Slater MSS, Spinning Mills, vol. 1, Spinning Mills to Almy and Brown, August 23, 1796; February 8 and March 20, 1797; January 16, 1800; see also Alexander Keyssar, "Widowhood in Eighteenth-Century Massachusetts: A Problem in the History of the Family," *Perspectives in American History* 8 (1974):99.

guarantee women any measure of independence or economic security.[45] The factory system, however, offered such a promise.

Widows readily placed their children in the factories, but a dilemma arose when jobs became available for adults; widows had to decide whether to accept industrial employment themselves. While few women entered the factory and tended machines, many nevertheless tried to effect a compromise between the new economic opportunities presented by the factory and contemporary customs regarding gender boundaries. Some widows became pickers for the factory master, and others left their homes and accepted jobs as full-time workers; yet the desire to find a compromise between societal expectations and economic necessity was almost universal. Widow Rebecca Cole and her family of four began working for Slater in the summer of 1795. During the early years of operation, the bleaching process was performed in a traditional manner. In a clearing near the factory, Slater had set into the ground a series of stakes, and Cole, with the help of several children (undoubtedly including several of her own), stretched the newly spun yarn across the stakes and kept the yarn wet. Over a period of time the sun bleached the yarn white.[46] For decades this family remained employed by Almy, Brown, and Slater. Over a thirteen-year period, Cole earned enough money to buy a house and lot worth $1,300.[47] Under the factory system she achieved a modicum of personal and economic independence without sacrificing responsibility for the training and discipline of her children. Thus she partly resolved the contradiction faced by women in her position. For indigent widows the factory system represented a new opportunity, an alternative to remarriage, public relief, and the charity of family and friends.

Unskilled male householders also found the factory attractive. As farm hands, teamsters, and casual day laborers, they worked in

45. Lyle Koehler, *A Search for Power: The Weaker Sex in Seventeenth-Century New England* (Urbana, 1980), pp. 108–129.

46. Almy and Brown MSS, Slater to Almy and Brown, September 18, 1795; Slater MSS, Spinning Mills to Almy and Brown, Account of Rebecca Cole, September 6–March 7, 1796; Kulik, "Beginnings of the Industrial Revolution in America," pp. 205–207.

47. Kulik, "Beginnings of the Industrial Revolution in America," pp. 205–206; see also Almy and Brown MSS, Slater to Almy and Brown, September 18, 1795.

an unstable labor market and could not predict from one season to another whether they would be able to support their families adequately. Economic downturns, poor harvests, and inclement weather raised the possibility that they, too, might have to depend on their local community for assistance, and that the family unit might be split apart.[48] Yet for male householders, the implications presented by the factory system challenged rather than broadened traditional gender roles. By hiring primarily children, by assuming responsibility for their discipline and education, and by allowing householders to shift for themselves, the factory master threatened the householder's position as provider, teacher, guide and protector of wife and children.

Responsibility for feeding, housing, and clothing the family, once divided among the members of the family, began to pass to the children. Although society recognized that "every member of a family should cordially endeavor to promote the common good," and although "the condition of many parents is such that they need the labor of their children to assist in sustaining the family," householders rejected outright the notion that boys and girls should become the mainstay of the family.[49] Yet this was precisely what was happening. The Arnold Benchley family typified the "new" factory family. From July 16 through October 3, 1792, Benchley's six children earned approximately £6 in the factory; during the same period, Benchley earned 3s.[50]

This abrupt break with the past was accompanied by others. Custom dictated that parents, and more especially the father, should "establish as soon as possible, an entire and absolute authority" over his children.[51] Indeed, "govern your child well; that is teach him the principles of obedience, the habit of bowing to duty, of subject-

48. Carville Earle and Ronald Hoffman, "The Foundation of the Modern Economy: Agriculture and the Costs of Labor in the United States and England, 1800–60," *American Historical Review* 85 (December 1980):1055–1094.

49. Enos Hitchcock, *Memoirs of the Bloomsgrove Family*, 2 vols. (Boston, 1790), 2:156; S. G. Goodrich, *Fireside Education* (New York, 1838), p. 79.

50. Almy and Brown MSS, Account of Labour: Children's Work, 1792. Perhaps Benchley was employed elsewhere, but this village offered few opportunities for unskilled workers. See Gilbane, "Social History of Samuel Slater's Pawtucket," pp. 185–192.

51. J. Witherspoon, *A Series of Letters on Education* (New York, 1797), p. 25.

ing his will to the authority of a guide," wrote S. G. Goodrich in his handbook titled *Fireside Education*.[52] Yet under the factory system, male householders could no longer fulfill their obligations as "the natural guardians of their children, . . . to control, restrain and direct the appetites of childhood."[53] Slater began to usurp these responsibilities by performing tasks once considered the duty of the father. As manager of the factory, he set the children to perform certain jobs and punished them for wrongdoing. Within the factory he commanded the deference, obedience, and respect due a father. He also took charge of their moral and educational training when he placed the children in the Sabbath school.

For seven days a week children were dominated by the factory system. Enos Hitchcock warned men that "the obligations between parent and child, are reciprocal; and if neglected by the former, it can hardly be expected that they will be fulfilled by the latter."[54] To rely principally on children for support, to surrender all responsibility for discipline and welfare to the factory master, threatened male pride and self-respect. Householders did not relinquish traditional prerogatives without a struggle. Between 1795 and 1800 they fought Slater for control over the factory floor and for the right to retain their traditional roles as guardians and guides.

Householders wanted the right to discipline their children while they were at work. They entered the factory frequently, spoke to their children while they were working, and argued with Slater over their punishment. On one occasion Slater complained, "Caleb Greene yesterday took all his children out of the mill except one who ret'd. whom I told I did not want except I had spinners. This morning Greene applyed to me and asked if I did not want his children again. I answered Yes, providing he would let them stay in future and do as they were bid and after a little cool conversation we parted."[55] Greene's children returned to work, but Slater was still not satisfied with the situation: "I am now very short handed have but 3 spinners. B. Earle's Boy is verry sick and but little pros-

---

52. Goodrich, *Fireside Education*, p. 85.
53. Hitchcock, *Memoirs of the Bloomsgrove Family*, 1:215.
54. Ibid., 1:82.
55. Almy and Brown MSS, Slater to Almy and Brown, October 4, 1796.

pect of his being better, Samuel Greene is lame and lazy. What are we going to do for help."[56] The struggle over the discipline of children affected the efficient operation of the factory.

Parents also invaded the factory whenever they believed their children were being exploited or abused. And there was cause for alarm. The partners often did not have the resources to pay laborers regularly in cash or in kind. In January 1796 Slater complained to Almy and Brown: "You must not expect much yarn until I am better supplyed with hands and money to pay them with—several are out of corn and I have not a single dollar to buy any for them."[57] Little was done, and again Slater petitioned Almy and Brown: please send

> a little money if not I must unavoidably stop the mill after this week. It is now going on four weeks since I recd 15–20. Can you immagine that upwards of 30 people can be supplyed with necessary articles that cannot be gotton short of cash with that. . . . Or, do not you immagine anything about it. This is the 3d and last time I means to write until a new supply is arrived.[58]

For days parents had demanded that some action be taken to satisfy basic needs. To placate them, Slater threatened Almy and Brown that he would shut down the mill or even sell off part of the stock or the machinery. He warned his partners that he could not "bear to have people come round me daily if sometimes hourly and saying I have no wood nor corn nor have not had any of several days. Can you expect my children to work if they have nothing to eat."[59] Parents became increasingly desperate.

Their anger was fanned by conditions in the mill. Light, heat, and safety precautions were at a minimum. Here Slater sided with the parents and urged his partners to supply the necessary funds to operate the factory safely. In November 1793 he complained that "the children are quivering this morning at seeing it snowy and cold and no stoves [here]."[60] The injury of a child prompted

56. Ibid., October 17, 1796.
57. Ibid., January 26, 1796.
58. Ibid., February 19, 1796.
59. Ibid.
60. Ibid., November 14, 1793, quoted in Conrad, "Evolution of Industrial Capitalism in Rhode Island," p. 107.

further concern. Slater warned his partners that "you call for yarn but think little about the means by which it is to be made such as children."[61]

Unable to protect their children, unable to force Almy, Brown, and Slater to provide safe working conditions, in desperate need of the food and wages promised to their children, parents took measures into their own hands. In 1795 they threatened to withdraw their sons and daughters from the mill unless the situation within the factory improved and Almy and Brown paid the children with more regularity. Almy and Brown ignored their pleas. In June of that year three householders struck the mill. In September and again in January and October of the following year, parents took their children out of the mill and effectively shut it down.[62]

Although stripped partially of economic power, the householder still dominated the family economy, and he retained considerable authority within it to discipline wife and children, to protect kin, to lead the family in prayer, and to supervise the educational and moral training of sons and daughters.[63] Men fought and children resisted any attempt by factory masters to encroach on these prerogatives. The British system of contract child wage laborers, like the apprenticeship program, would not work well in the United States.

To recruit and retain a supply of tractable, steady workers, Slater had to effect some sort of compromise with householders whereby their customary values and their social and economic position within the family and the wider society would be safeguarded, and they would be able to retain their dignity and self-respect. Slater recognized the claims of householders as legitimate and took measures that strengthened patriarchy among the lower classes. He introduced the family system of labor; by which the entire family was brought under the umbrella of the factory system. A division of labor based on age, gender, and marital status was put into place. While children tended machines, their fathers were promised so-

61. Almy and Brown MSS, Slater to Almy and Brown, December 24, 1794, quoted in Conrad, "Evolution of Industrial Capitalism in Rhode Island," p. 108.

62. Almy and Brown MSS, Slater to Almy and Brown, June 2 and September 24, 1795; January 7, October 4, and November 10, 1796.

63. Alexis de Tocqueville, *Democracy in America*, ed. J. P. Mayer, trans. George Lawrence (Garden City, 1969 [1966]), pp. 600–603.

cially acceptable forms of labor. At the turn of the nineteenth century this meant work outside the factory walls as construction workers, farm hands, teamsters, watchmen, and groundskeepers. By providing men with such jobs and by paying them adult male wages, Slater reinforced the male householder's position with the family as primary provider. Within the factory context, he made other concessions. Slater allowed householders to share responsibility over the allocation of jobs, the training of children, and the discipline of workers. This system bolstered the householder's role as guardian, guide, and teacher. The relationship between Slater and the householder was a reciprocal one, as householders provided Slater with a steady supply of tractable hands.

Slater's program, however, could not be developed fully in Pawtucket. The community and surrounding area offered few acceptable economic opportunities for unskilled men. Besides, Almy and Brown placed obstacles in Slater's way. Ignorant of factory operations, unaccustomed to handling large numbers of people drawn from the lower ranks of society, removed from daily contact with workers and their problems, and preoccupied with costs, these merchants sabotaged Slater's labor policies. Like his Pawtucket partners, by 1799 Samuel Slater believed it was time to establish his own factory where he could introduce the ownership, management, and labor programs he felt would prove efficient and successful.

The years spent with Almy and Brown nevertheless proved invaluable for Slater. During this period he learned through trial and error what forms of ownership, management, and labor operated best in the United States. British concepts could not be transplanted easily; they had to be modified to meet the demands of the local population. What emerged by 1800 was a factory system uniquely American, an extension of the values, institutions, and ideas present in New England society. And a chief characteristic of this system was patriarchy. It could be observed in the family firm, where kin both owned and managed the business, and in the family system of labor that emerged in the factories. All rested firmly upon traditional, eighteenth-century familial values. Slater realized the strength of patriarchy in America and organized his factory system to accommodate it.

PART II

THE FORCE OF
TRADITION

[4]

Forms of Ownership and Management

From Pawtucket the factory system spread throughout Rhode Island, Connecticut, Massachusetts, Pennsylvania, and New York. At first, growth was slow. It has been estimated that by 1808 only fifteen spinning mills were operating in the United States and that almost half of them belonged to Samuel Slater, William Almy and Obadiah Brown, or one of Slater's former employees.[1] Some manufacturers and their associates feared that the local market could not absorb any increase in yarn production. Writing to his children in the fall of 1810, Moses Brown urged caution in the expansion of their business: "Our people have 'cotton mill fever' as it is called. Every place almost occupied with cotton mills; many villages built up within 16 miles of town and spinning yarn and making cloth is become our greatest business. We were first to get into it. Samuel Slater has sold out one half of one mill and I should be pleased my children could do, with their four, in part as he has done."[2] While Brown exaggerated the situation, his concern had merit. British competition, a limited national market for domestic yarns, and, of course, increasing competition from other manufacturers were considerations. But none of these factors curbed the "cotton mill fever."

1. Albert Gallatin, "Manufactures," April 17, 1810, in *American State Papers, Finance*, 2:427; Peter J. Coleman, *The Transformation of Rhode Island, 1790–1860* (Providence: 1963), pp. viii, 130–131; Ware, *Early New England Cotton Manufacture*, pp. 30, 128, 301–302.
2. Almy and Brown MSS, Moses Brown to T. Rogerson, November 11, 1810.

When the Embargo and the War of 1812 effectively cut off the supply of British yarn and cloth goods, the American industry boomed. Potentially profitable outlets for capital were scarce during the war years, and merchants, traders, and professional men began to invest in the emerging textile industry. One of the most innovative and enterprising men to enter the industry was Francis Cabot Lowell, a Boston merchant. He introduced an industrial system that was characterized by innovative technology, the integration of spinning and weaving, corporate ownership, professional management, and the use of a female labor force. Rather than manufacture yarn exclusively, Lowell perfected a power loom and produced sturdy, inexpensive coarse cloth for the Western trade. Successful and especially profitable, this system achieved worldwide recognition for its novel approach to industrialization. Historians have claimed that the Lowell system "was the most important thing which could have happened to the cotton industry," for it was "taken up by men with the best business imagination in the land, unhampered by its traditions, concerned with making fortunes and building states, not with manufacturing cotton cloth."[3] Yet in 1813 and 1814 the full implications of the Lowell system remained to be seen. Lowell was only one of many men who took advantage of the temporary dislocation in the textile industry and tried to capture a part of the local market.[4] By 1814 Tench Coxe estimated that 243 cotton mills operated within fifteen states; Pennsylvania, Massachusetts, Rhode Island, and New York led the way with 64, 54, 28, and 26 mills, respectively.[5]

---

3. Ware, *Early New England Cotton Manufacture*, pp. 61–62.

4. The Lowell system has been researched well. Among the better books and articles are Ware, *Early New England Cotton Manufacture*; Hannah Josephson, *Golden Threads: New England's Mill Girls and Magnates* (New York, 1949); Nathan Appleton, *Introduction of the Power Loom, and Origins of Lowell* (Lowell, 1858); Dublin, *Women at Work*; Kenneth F. Mailloux, "The Boston Manufacturing Company," *Textile History Review* 5 (January 1964): 3–29; Carl Gersuny, " 'A Devil in Petticoats' and Just Cause: Patterns of Punishment in Two New England Textile Factories," *Business History Review* 50 (Summer 1976): 131–152. See also *Niles Weekly Register*, June 3 and September 9, 1826; May 19, 1827; July 5 and July 26, 1828; July 6, 1833; Louis McLane, "Report of the Secretary of the Treasury," 1832, *Documents Relative to the Manufactures in the United States*, 2 vols., 22d Cong., 1st sess., House Executive Document no. 308 (Washington, 1833), 1:340–341 (hereafter referred to as *McLane Report*); "Statistics of Lowell, Massachusetts," *Hunts Merchants Magazine* 1 (July–December 1839):90.

5. Tench Coxe, "Digest of Manufactures," in *American State Papers, Finance*, 2:666–812.

At the war's end, however, the boom collapsed and British goods flooded the market; many American firms shut down. The *Niles Weekly Register* reported in November 1816 that "the importation of British goods is yet enormous, and they are selling at prices insufficient to pay costs and charges—the pound sterling of the invoice is often, it is said, fairly sold for the pound currency at New York; and all, or nearly all, of our large manufacturing establishments have more or less suspended business."[6]

Costs had to be cut. Robert Zevin believed that it was during this crisis that American textile manufacturers turned increasingly to the power loom. Two looms were on the market; one made at Waltham sold for about $125, and a second machine, developed by William Gilmore, cost $70. No patent limited the spread of the Gilmore loom, and soon owners of the Lyman Mills in North Providence and the proprietors of the Coventry factory several miles away introduced a version of this machine. Dexter Wheeler, who was associated with the Fall River Manufacturing Company, copied this pattern, and the machine spread throughout southern New England. The Waltham and the Gilmore looms were immensely beneficial to the textile industry. Savings of 4, 6, and 9 cents a yard could be achieved through the introduction of a power loom. Many mills that formerly had been engaged in yarn production switched to the new loom and combined spinning and weaving in their factories.[7] In Massachusetts, Rhode Island, and Connecticut, integrated factories with power looms numbered eleven, fifteen, and eleven, respectively, by 1820.[8] Cloth output soared: Zevin estimated that between 1816 and 1833 cloth production increased at a compound annual rate of 39 percent.[9]

6. *Niles Weekly Register*, November 16, 1816.

7. Robert B. Zevin, *The Growth of Manufacturing in Early-Nineteenth-Century New England* (New York, 1975), pp. 10–3, 10–4, 10–12.

8. Peter J. Coleman, "Rhode Island Cotton Manufacturing: A Study in Economic Conservatism," *Rhode Island History* 23 (July 1964): 68; Bagnall, *Textile Industries of the United States*, p. 546; David John Jeremy, *Transatlantic Industrial Revolution: The Diffusion of Textile Technologies between Britain and America, 1790–1830s* (Cambridge, Mass., 1981), p. 98. Thomas R. Smith, *Cotton Textile Industry of Fall River, Massachusetts: A Study in Industrial Localization* (New York, 1944), p. 24; see also an advertisement for the Gilmore loom in *Manufacturers' and Farmers' Journal and Providence and Pawtucket Advertiser*, April 6, 1820; and Slater MSS, Samuel Slater and Sons, vol. 191.

9. Zevin, *Growth of Manufacturing in Early-Nineteenth-Century New England*, p. 8.

Further reductions in the cost of production were achieved as the price of raw cotton declined throughout the era. For upland cotton, which sold in the New York market, manufacturers paid approximately 31 cents a pound in 1818 and about 10 cents a pound in 1832.[10] Cloth prices reflected these changes. Prices charged for goods manufactured by the Boston Manufacturing Company, for example, were almost cut in half between 1823 and 1833. Other manufacturers had to scramble to keep pace with these prices; they tried to cut costs where they could.[11]

FACTORIES AND FIRMS

Samuel Slater participated in the phenomenal growth of the industry. His career was an eminently prosperous one, and at his death in 1835 he was operating in three states. At one time he either owned or held an interest in at least thirteen textile mills, two machine shops, and a wholesale and commission firm (see Table 2). Slater's factories and firms served as the models for hundreds of would-be manufacturers throughout the United States.

Slater formed his first factory independent of Almy and Brown in 1799, the year his Providence partners bought the Warwick Spinning firm. With the financial support of Oziel Wilkinson, William Wilkinson, and Timothy Green, all related to him through marriage, Slater formed Samuel Slater and Company and built the "White Mill" on the Massachusetts side of the Pawtucket River, at Rehoboth. This two-story spinning factory measuring 49' by 26' resembled the "old mill" at Pawtucket in size, appearance, and the type of goods manufactured. Slater continued in partnership with Almy and Brown. For years he shuttled between Pawtucket and Rehoboth to supervise operations at both mills.[12] From the outset

10. Arthur H. Cole, *Wholesale Commodity Prices in the United States, 1700–1861: Statistical Supplement: Actual Wholesale Prices of Various Commodites* (Cambridge, Mass., 1938), pp. 184–242.
11. Zevin, *Growth of Manufacturing in Early-Nineteenth-Century New England*, pp. 10-1-10-49; see also Ware, *Early New England Cotton Manufacture*, p. 111.

Table 2. Firms founded, purchased, or partially owned by Samuel Slater, 1790–1835

Firm	Date acquired
Almy, Brown, and Slater	1790
Samuel Slater and Company	1799
Almy, Brown, and the Slaters	1806
Slater and Tiffany*	1811
Slater and Howard†	1815
Providence Iron Foundry	1815
Jewett City Factory	1823
Phoenix Thread Company*	1824
Amoskeag Manufacturing Company	1825
Slater and Wardwell	1827
Slater and Kimball*	1827
Steam Cotton Manufacturing Company	1827
Central Falls Mill	1829
Sutton Manufacturing Company	1832
S. & J. Slater	1832
Providence Machine Company	1834

*Succeeded by Union Mills.
†Succeeded by Dudley Manufacturing Company and later by Webster Woolen Company.

the two firms were run as one. This decision received the whole-hearted support of Slater's several partners. As Moses Brown explained: "You may well think that the Erection of that mill [Samuel Slater and Company] was not agreeable to my Ideas but after it was determined on . . . in order to save the business from Immediate ruin We thought best, to so far unite so as not to interfere with each other in Workmen nor wages."[13] This attitude carried over into the prices proprietors charged for goods. To New York agent Gilbert Everingham, Almy and Brown wrote: "Slater is connected with us in the business & we unite in our prices of yarn &

12. Gilbane, "Social History of Samuel Slater's Pawtucket," pp. 136–138, 146–147. In January 1810 Slater dissolved his partnership with his kinsmen. For an excellent discussion of Slater's various business transactions see N. S. B. Gras and Henrietta M. Larson, *Casebook in American Business History* (New York, 1939), pp. 218–229; La Porte, "Birth of America's Spinning Industry, II," pp. 678–679; Orra L. Stone, *History of Massachusetts Industries: Their Inception, Growth, and Success*, 4 vols. (Boston, 1930), 1:38–39; Samuel Batchelder, *Introduction and Early Progress of the Cotton Manufacture in the United States* (Boston, 1863), p. 49.
13. Almy and Brown MSS, Moses Brown to E. Waterman, February 23, 1802, quoted in Conrad, "Evolution of Industrial Capitalism in Rhode Island," p. 157.

they are exactly conformable to theirs."[14] It should be noted also that the firms pooled their resources when they purchased cotton. The Rehoboth factory was only one of Slater's many successful business ventures.[15]

In 1806 Almy, Brown, and Slater settled some of their outstanding differences, and the three partners joined forces with John Slater, Samuel Slater's brother, to purchase 122 acres of land on the Mohegan River in upstate Rhode Island and construct another yarn-spinning mill. John Slater managed the factory and planned the community of Slatersville, which developed around it. The Slater brothers continued to invest in the region, and eventually they purchased Almy and Brown's share in the enterprise. By the 1830s S. & J. Slater, as their firm was called, ran at least four factories, operated 9,500 spindles, and provided employment for 66 men, 109 women, and 169 children. In addition to the factories, the outbuildings, and the company cottages, the Slater brothers acquired more than 1,200 acres of land in the Slatersville area.[16] Throughout the period the management of Slatersville remained the responsibility of John Slater.

By the first decade of the nineteenth century, Samuel Slater held part ownership in three factories in Massachusetts and Rhode Island. With management of the Pawtucket and the Rehoboth factories weighing him down, he sold his interest in the Rehoboth factory in 1810 and invested immediately in another Massachusetts concern.[17]

Samuel Slater wanted to strike off on his own. He entered into a partnership in 1811 with Bela Tiffany, a former employee and a family friend, to purchase almost 270 acres of land and water priv-

14. Almy and Brown MSS, Almy and Brown to Gilbert Everingham, October 2, 1802, quoted in Conrad, "Evolution of Industrial Capitalism in Rhode Island," p. 158.
15. Ibid.
16. White, *Memoir of Samuel Slater*, pp. 215, 259; *McLane Report* 1:970–971; Bagnall, *Textile Industries of the United States*, pp. 397–400; James Montgomery, *Practical Detail of the Cotton Manufacture of the United States of America and the State of the Cotton Manufacture of That Country Contrasted and Compared with That of Great Britain with Comparative Estimates of the Cost of Manufacturing in Both Countries* (Glasgow, 1840), pp. 153, 179; Batchelder, *Introduction and Early Progress of the Cotton Manufacture*, pp. 52–53; Lewton, "Samuel Slater and the Oldest Cotton Machinery in America," p. 507.
17. La Porte, "Birth of America's Spinning Industry, II," p. 679.

ileges in Oxford Township, in south-central Massachusetts. Slater and Tiffany probably began operations a year later.[18] Although Tiffany, who held a one-sixth share in the enterprise, managed this factory, Slater retained a keen interest in the venture. Independently of his partner, he increased his holdings in the region to include a woolen mill, a dye and bleach house, a saw- and gristmill, sixteen dwellings, and 700 acres of land.[19] Of these investments the woolen factory was the most important. With fellow Englishman and friend Edward Howard, Slater built a small woolen mill near the Slater and Tiffany factory in 1815 and rebuilt it after it was destroyed by fire in 1820. A notice published in the *Massachusetts Spy* on January 15, 1823, declared the new factory open for business:

> The Subscribers hereby give notice, that they have formed a Co-partnership, for the purpose of manufacturing Woollen Goods, under the Firm of SLATER & HOWARD.
>
> The business will be carried on as usual, in all its branches.[20]

Initially, Howard managed the business. The Slater and Tiffany mill together with the Slater and Howard factory formed the nucleus of a new community. The area was known as Oxford South Gore until it was incorporated as Webster in 1832. To avoid confusion, the name Webster will be used to designate this factory colony throughout the remainder of this book.

Next Slater turned his attention to Connecticut. He formed a second partnership with his brother in 1823, and together they

18. "Two Hundred Years of Progress: Webster-Dudley, 1739–1939," *Webster Times,* 1939, p. 7; Slater and Sons, *Slater Mills at Webster,* p. 25; *Massachusetts Spy* (Worcester), January 15, 1823; Lewton, "Samuel Slater and the Oldest Cotton Machinery in America," p. 507.

19. "Two Hundred Years of Progress," pp. 8–9; Slater and Sons, *Slater Mills at Webster,* pp. 22, 25; George Robert Means, "The Industrial Development of Webster, Massachusetts," Master's thesis, Clark University, 1932, pp. 53, 68–70; D. Hamilton Hurd, *History of Worcester County, Massachusetts, with Biographical Sketches of Many of Its Pioneers and Prominent Men,* 2 vols. (Philadelphia, 1889), 1:362; Lewton, "Samuel Slater and the Oldest Cotton Machinery in America," p. 507.

20. *Massachusetts Spy* (Worcester), January 15, 1823; see also Slater and Sons, *Slater Mills at Webster,* p. 25; James Lawson Conrad, Jr., "The Establishment of the Textile Industry in Dudley, Massachusetts, 1812–1920," Master's thesis, Clark University, 1963, p. 56.

purchased for $17,000 a small mill in Jewett City, New London County. Once again management fell to John Slater, who built up the site by constructing a small woolen factory, a saw- and gristmill, and several dwellings. While the area was being developed, Samuel Slater remained in the background, and in 1831 he sold his interest to his brother.[21]

Slater continued to expand. Interest in a New Hampshire firm, the celebrated Amoskeag Manufacturing Company, coincided with his involvement in the Jewett City venture. Although inquiries were initiated earlier, by 1825 it appeared that Slater, together with Oliver Dean, Lyman Tiffany, Willard Sayles, Larned Pitcher, and Ira Gay, had secured property and an unfinished mill in Manchester, New Hampshire. But Slater's involvement in this factory was negligible, and his interest and activities shifted again to the southern part of the region.[22] Between 1827 and 1831 Slater built, bought outright, or acquired an interest in several other factories, including the Central Falls Mill in Smithfield, the Sutton Manufacturing Company in Wilkinsonville, Massachusetts, and the Steam Cotton Manufacturing Company in Providence, Rhode Island. Of the three factories, the Steam Cotton Manufacturing Company was the most noteworthy because of the novel mode of power employed there. Steam engines operated the 4,344 spindles, 18 mules, and 100 new power looms the mill used to manufacture fine-quality sheeting and shirting. The size and scale of the new factory can be estimated from entries in the company's daybooks concerning its construction. Built of stone and plated with brick, this two-story factory, measuring 30' by 43', was similar in size to Slater's earlier mills in Pawtucket and Rehoboth. A stone furnace was added to it soon after the factory was built.[23] This mill was one of three in Rhode Island in the 1820s and early 1830s to use steam power.[24]

Slater's business enterprise was not confined to textile mills. In

21. Bagnall, *Textile Industries of the United States*, pp. 595–596.

22. William Davis, ed., *New England States: Their Constitutional, Judicial, Educational, Commercial, Professional, and Industrial History*, 2 vols. (Boston, 1897), 1:150.

23. Slater MSS, Steam Cotton Manufacturing Company, vol. 11, especially 1827 and October 6, 1834; see also Leander J. Bishop, *History of American Manufactures from 1608 to 1860*, 3 vols. (Philadelphia, 1864), 3:387–388.

24. Coleman, *Transformation of Rhode Island*, p. 109; Alfred D. Chandler, Jr., "Samuel Slater, Francis Cabot Lowell, and the Beginning of the Factory System in the United States" (Harvard Business School, 1977) p. 23 (mimeo).

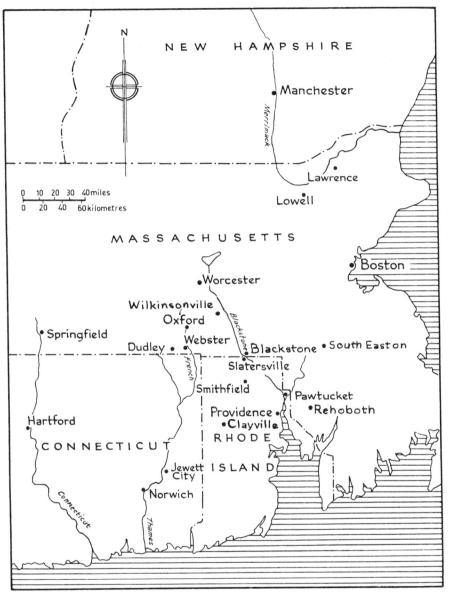

Southern New England

[97]

the 1820s he began to diversify. To circumvent the customary commission house system and thereby to avoid the fees paid to wholesalers, shopkeepers, and other agents, Slater purchased an interest in George Wardwell's Providence wholesale firm and instructed Wardwell to sell part of his cloth, yarn, and thread through this outlet. Although Wardwell objected to the scheme and argued that "after making particular inquirie respecting the sale of woolen goods at this place . . . I think it best for us to relinquish the idea of attempting the business at present," Slater persisted with his plans.[25] Within a month Slater and Wardwell sold not only woolen goods but also satinets and cotton goods of all descriptions. But Wardwell lacked the knowledge to market such competitive products successfully. In the fine and fancy goods market, British, not American, manufacturers determined "what *is* and what *is to be* 'the go.'"[26] Style, fashion, and color followed British and European trends, and American manufacturers had to keep pace with the latest changes if they hoped to compete successfully. When Wardwell proved to be incapable, Slater reverted to commission agents in New York, Boston, New Orleans, Philadelphia, Baltimore, Providence, and Hartford to sell his cloth. But the Slater and Wardwell firm had set a precedent. This was the first but not the last time Slater and, later, his sons would try to sell their products without the aid of middlemen.[27]

Diversification was not limited to marketing. When Samuel Slater built the Steam Cotton Manufacturing Company in 1827, he organized a machine shop on the first floor of the factory and hired Thomas J. Hill to take charge of it. Four years later the two men entered a partnership and formed a new firm, the Providence Machine Company. This enterprise supplied not only the Slater factories but also manufacturers throughout New England with dou-

25. Slater MSS, Slater and Wardwell, vol. 10, Wardwell to Slater, February 10, 1829.
26. Ibid., Samuel Slater and Sons, vol. 216, Tiffany, Anderson Company to Samuel Slater and Sons, October 12, 1833.
27. Ibid., Union Mills, vol. 117, Storrs to George Blackburn and Company, February 13, 1845; Storrs to R. and D. M. Stebbins, February 28, 1845; Samuel Slater and Sons, vol. 210, Underhill and Company to Samuel Slater and Sons, July 3, 1845; see also Slater and Sons, *Slater Mills at Webster*, pp. 33–34.

blers, mules, spreaders, stretchers, dressers, and lappers.[28] Most of the iron used in the construction of the equipment was purchased from another Slater venture, the Providence Iron Foundry, a partnership formed with his brother-in-law, David Wilkinson, and five other men. In 1847 Hill purchased the Slater interest in the machine company for almost $50,000 and relocated the business.[29]

THE SLATER SYSTEM

The Slater factories shared certain features: the type of product manufactured, the technology employed, the partnership form of ownership, and personal management. These characteristics became the hallmark of a model of production which was adopted by hundreds of manufacturers throughout the country. During his life Samuel Slater remained at the vanguard of the system that came to bear his name.

Throughout much of his career, Slater confined the products of his spinning factories to fine yarn, thread, and twine. In this emphasis on yarn spinning, Slater differed from many of his contemporaries. He was also unlike many other factory owners in his slowness to introduce the Gilmore power loom. Slater ignored the mechanical device until 1823, when he installed it in the cotton mills; later he mechanized the woolen branch of his business. Although the cotton power loom had been adapted successfully to woolen cloth weaving following the War of 1812, Samuel Slater did not introduce this equipment until 1829–30. The delay cost Slater dearly.[30]

---

28. Slater and Sons, *Slater Mills at Webster*, p. 25; "Two Hundred Years of Progress," pp. 8–9; Slater MSS, vol. 235, Samuel Slater to John Slater, April 22, 1831; Steam Cotton Manufacturing Company, vol. 11, February 18, 1831; April 28, and December 15, 1830; February 10, March 28, and May 13, 1831; January 13, 1832; May 6, 1833; see also Bishop, *History of American Manufactures*, 3:387–388.

29. Bishop, *History of American Manufactures*, 3:287–288; Coleman, *Transformation of Rhode Island*, pp. 150, 111.

30. Coleman, "Rhode Island Cotton Manufacturing," p. 68; Bagnall, *Textile Industries of the United States*, p. 546; Batchelder, *Introduction and Early Progress of the Cotton Manufacture*, pp. 70–72; *Manufacturers' and Farmers' Journal and Providence and Pawtucket Advertiser*, April 6, 1820.

During the transition in the woolen mills, Slater kept a ledger of the costs involved in weaving woolen goods both by hand and by the new loom. A comparison of the costs showed vividly the savings that could be effected by the new machine. In August 1829 Samuel Slater employed twenty-two male hand-loom weavers at his Webster woolen factory, and he paid them between $0.14 and $0.20 a yard for weaving various types of broadcloth and $0.06 a yard for weaving kersey. Those who worked steadily earned between $24.00 and $30.00 that month. William Archer, for example, wove slightly more than 201 yards of steel mountain, claret, and olive broadcloth at $0.14 a yard for a total income of $28.21. Earlier in the year the firm had installed power looms, had employed young women to operate them, and had begun to manufacture various types of broadcloth. By August of 1829, thirteen women worked in the new weaving room, and they received $0.08 per yard for their cloth. A woman operating a power loom could produce roughly the same amount of cloth as a hand-loom weaver. The company records over the next ten months show a gradual decline in the number of hand-loom weavers employed at this factory. By June 1830 only six of them remained on the payroll. While they continued to receive high piece rates of between $0.12 and $0.20 a yard for their cloth, they worked fewer and fewer hours each month. During the same period, the number of power-loom operators rose to sixteen, and they worked steadily. All received $0.06½ a yard for their fabric. By adopting the power loom, the Slater family had cut weaving costs by one-half or more.[31]

Slater's reluctance to adopt the new equipment was not due to the cost of conversion; expenses were not exorbitant. Explanations lie elsewhere. Workers could have opposed the introduction of new technology and delayed its adoption. The power loom was operated largely by single, itinerant women, and Slater employed entire families in his mills. To recruit workers he would have had to look beyond the kinship network. Or Slater may have been content not to develop his market further. By the 1820s Slater's name was synonymous with quality yarn, thread, and hand-woven cloth, and he

31. Slater MSS, Samuel Slater and Sons, vol. 191, July 1829–July 1830; see also Zevin, *Growth of Manufacturing in Early-Nineteenth-Century New England,* pp. 10-41, 10-42.

shared the market for these goods in part with British producers. Or perhaps Samuel Slater was less cost-conscious than his competitors. Committed to a system of cloth production which had served him well for decades and which provided employment for hundreds of people in and near his factory colonies, he saw few reasons to change. Wholesale adoption of the power loom occurred when his sons assumed control of the family business in 1829.[32]

Technological innovation proceeded slowly in other areas as well. With the exception of the Steam Cotton Manufacturing Company, almost all of Slater's firms relied primarily on water power to operate equipment. His factories were located in rural areas of Rhode Island, Massachusetts, and Connecticut where water resources were available. His preference for water power could be seen in his largest and most highly developed factory colony, located in Webster, Massachusetts. There in 1812 Slater constructed his first factory near Chaubunagungamaug Pond, a large body of water that empties into the French River. From its origin near Leicester, Massachusetts, the French River drains an area of fifty square miles above this region. This drainage basin, together with the natural ponds and later the artificial reservoir built to hold back the water and to regulate its flow throughout the year, made the Oxford South Gore area an ideal site for water-powered factories. All of Slater's factories there were powered by huge breast-type water wheels.[33] In the initial stage of factory operations at Webster, sufficient water flow allowed for easy expansion of the factories. Water power also was cleaner and probably was cheaper than steam power. Expenses for steam power included the purchase of coal, transportation of the coal to an inland location, and the employment of a person to supervise the engine and the boiler room.[34] In most of the Slater mills, steam engines were not introduced until the Civil War, and then they served primarily as a reserve, emergency power source and not as the main mode of power.[35]

32. Workers' opposition to the new technology is discussed in chap. 9.
33. Means, "Industrial Development of Webster," pp. 12, 17, 18, 58.
34. Louis C. Hunter, *A History of Industrial Power in the United States, 1780–1930* (Charlottesville, 1979), vol. 1, *Waterpower in the Century of the Steam Engine*, 519–520.
35. Until the 1850s water power was the primary mode of power throughout the New England textile districts. Manufacturers only reluctantly abandoned water wheels. Severe climatic conditions such as drought encouraged some manufacturers to in-

In all of Slater's factories, ownership and management went hand in hand. Adopted first at Pawtucket, the partnership form of ownership was retained by Slater despite the advantages of incorporation, an ownership form that had become popular during the War of 1812. If granted a special incorporation charter by the state, firms were guaranteed perpetual life, limited liability, and the right to accumulate capital. Through their charters, businessmen also received special concessions, monetary grants, and other privileges. By the war's end, seventy-five textile firms had received state charters, but Slater's was not among them.[36] He was not persuaded to adopt this form of ownership; he preferred traditional partnership agreements or single proprietorships.

For Slater's enterprises capital requirements were limited: one or two partners usually supplied the funds necessary to construct a factory. By reinvesting his profits, Slater could expand operations.[37] This was common practice throughout much of the industry. Zachariah Allen, for example, rejected offers of outside capital when he established his factory and wanted to expand. In a letter to Francis C. Lowell II, a prospective investor, he explained: "I commenced my works about four years since with the intention of extending them no farther than my own capital would allow without inconvenience, I still think it would be preferable to continue in the course I have adopted."[38]

In the selection of suitable partners, only kin, close family friends, and business associates—all considered trustworthy individuals—proved acceptable to Slater. David Wilkinson, a partner in several Slater enterprises, was his brother-in-law; John Slater, his Slatersville and Jewett City partner, was his brother; Thomas Hill and

---

stall steam engines. This was the case in Waltham, where in 1836 the owners of the Boston Manufacturing Company installed a steam engine as an auxiliary source of power to be used during periods of low water. Other manufacturers switched from water to steam power when they expanded operations and found that local streams could not supply adequate power. See Hunter, *History of Industrial Power in the United States*, 1:9–10, 516–521.

36. Coleman, *Transformation of Rhode Island*, p. 109; Chandler, "Samuel Slater, Francis Cabot Lowell, and the Beginning of the Factory System," p. 23; see also Alfred D. Chandler, Jr., "A Reply," *Business History Review*, 53 (Summer 1979):256.

37. *McLane Report*, 1:576–577.

38. Zachariah Allen to F. C. Lowell, November 30, 1825, Francis Cabot Lowell II, Business Correspondence, 1793–1827, Massachusetts Historical Society, Boston (hereafter referred to as Lowell MSS).

Bela Tiffany, respectively partners in a machine shop and a cotton mill, were former employees; George Wardwell and Edward Howard, both close family friends, were partners in a commission business and in several factories. Much of his property came to be held by Samuel Slater and Sons, a closed family partnership formed with three of his sons, George, John II, and Horatio Nelson Slater.[39]

Slater's preference for the partnership form of ownership probably had more to do with his attitude toward management than with concern over costs. With the partnership form of ownership, a simple line of organization could be adopted to manage the firms: proprietors not only owned the concern but also usually managed it, often working alongside laborers on the factory floor. Like his Pawtucket partners, Slater came to distrust outsiders and believed that a business failed when its owners, "who themselves engaged in other pursuits, have invested the net profits of their business in manufacturing and left the latter to the superintendence of others," for "it is in this triple capacity of money lender, employer and laborer that our most successful manufacturers have succeeded."[40] He preferred to superintend the factories himself or to entrust the task to one of his partners or sons.

Slater's management policies worked well for a number of years. Problems arose, however, when the partners wanted to resign and leave the firm. Where could Slater secure a competent, trustworthy, knowledgeable supervisor? Not willing to entrust responsibility to paid professional agents, he turned to his family for assistance, and he increasingly drew his sons into his business activities. The factories at Webster became a training ground for his children, who one day would control the family business. For his sons, this was not an easy apprenticeship.

FAMILY MANAGEMENT

Although Slater's interest in Webster dated from the War of 1812, it was not until the end of the decade that Slater began to focus attention on the area. Beginning in 1818 he acquired full owner-

39. Slater and Sons, *Slater Mills at Webster*, p. 25.
40. *McLane Report*, 1:929.

ship of the Slater and Tiffany factory, bought the Slater and Howard mill, and then built a third factory, the Phoenix Thread Mill.
Repeatedly these factories were enlarged, first in 1822, then in 1826
and 1827, and again in 1828.[41] His early partnerships with Tiffany
and Howard had been a necessary expedient that had allowed him
to limit capital outlay, to secure the services of a factory manager,
and to see if local labor resources, transportation networks, and water
power could support large-scale production. Convinced that the
area had potential, he began to acquire ownership of the various
enterprises in the region, but these acquisitions brought problems. With the retirement of Tiffany and later of Howard, Samuel
Slater assumed responsibility for managing the mills. He still supervised directly the old mill at Pawtucket and tried to assist his
brother in the management of the Slatersville concerns. He found
he could not do it all. Traveling the almost seventy miles between
Pawtucket and Webster taxed his strength, and supervision of the
Webster property presented special problems. At Pawtucket the division of labor initiated among the partners relieved Slater of responsibility for the purchase of supplies, the marketing of manufactured goods, and the supervision of the putting-out system: Almy
and Brown managed these operations and left Slater in charge of
the factory. This was not the case at Webster, where Slater had to
oversee the factory, recruit and discipline an industrial labor force,
plan and design a community for his workers, purchase supplies
and sell products, and supervise a vast putting-out network.

In the early years at Webster Slater employed more than five
hundred part-time hand-loom weavers. While some of the weavers
were the parents of the children who worked in his factory and
lived nearby, most of them lived in the surrounding towns of
Thompson, Dudley, Oxford, and Charlton and periodically traveled to the Slater mills to collect materials. When Samuel Slater
dispensed yarn, he provided outworkers with detailed instructions
on how the cloth was to be woven. In part, the weaver's ticket read:
"Weavers must return the Yarn left of a piece, with the cloth—
Cloth must be trimmed, wove as thick at the ends as in the middle,
and returned free from stains and dirt and if it is made too sleazy,

41. Ibid., 1:576–577.

or damaged in any way, a deduction will be made from the weaving."[42] The threat was not an idle one, for Slater examined thoroughly every piece of cloth. If weavers took more than four months to complete their work, they were docked a half-cent a yard on the cloth returned, and if they failed to return all of the yarn given out, they were charged for it and dismissed.[43] This procedure proved burdensome, to him, and in the early 1820s Slater began to employ subconstractors to handle it.

"Merchant Weavers attend!" began a notice placed in the *Massachusetts Spy* (Worcester) on January 8, 1823. "A few yards good YARN will be furnished to WEAVE on reasonable terms, on application to the Subscriber at his Factory in Oxford [Webster]."[44] To men who answered this advertisement, usually shopkeepers who operated putting-out systems as sidelines to their regular retail trade, Slater transferred the burden of finding weavers and supervising their work. Merchant weavers now collected yarn from the factory, distributed it to country weavers, checked their work for quality and regularity of weave, paid them in cash or, more frequently, in store goods, and then returned the finished cloth to Slater.[45] Several years later Slater went a step further and converted to the power loom and thus integrated his Webster mills. Yet even with this modification, Slater could not manage easily both the Pawtucket and the Webster factories. Rather than seek another partner or hire a professional manager to operate one of them, Samuel Slater turned to his family for assistance. He was determined that his sons should follow him into the family business.

In 1791 Samuel Slater had married Hannah Wilkinson, daughter of a Pawtucket iron manufacturer, and they had nine children, two girls and seven boys. Eight of the children survived infancy.[46] Shortly after Tiffany's retirement in 1818, Slater employed the

42. Slater MSS, Slater and Tiffany, vol. 124.

43. Ibid., Slater and Tiffany, vol. 80, Hand-Loom Weavers, 1812–1829, especially Timothy Corbin, April–March 1819. See also Chandler, *Visible Hand*, p. 63; Gras and Larson, *Casebook in American Business History*, pp. 226–227.

44. *Massachusetts Spy* (Worcester), January 8, 1823.

45. Coleman, "Rhode Island Cotton Manufacturing," pp. 71–77; Ware, *Early New England Cotton Manufacture*, p. 74; Slater MSS, Union Mills, vol. 169, Wage receipts, Artimas See and Company, June 15, 1827; George Bauen, August 10, 1827; Samuel Slater and Sons, vol. 191, Hand Loom Weavers.

46. White, *Memoir of Samuel Slater*, pp. 241–242.

adolescent Samuel Slater II, but the boy was frail, and he died. Effective supervision of the Webster property once again reverted to Samuel Slater, who subsequently wrote to his son John Slater II, recently turned sixteen: "You will have to make your appearance at Oxford [Webster] or here in one of the stores as per conversation with you some time past. It is highly important that one or more of my sons was learning the business so as to in some measure relieve me from the close attention which I have to attend."[47] The boy was taken out of school and sent to the old "Green Mill." John Slater II served a long apprenticeship under his father. Although Samuel Slater was fifty-three years of age and suffered from rheumatism, he did not allow his son appreciable discretion in the actual operation of the factory. Through numerous personal visits to Webster and frequent detailed correspondence, he supervised John and, indirectly, the factories. In March 1826, for example, he wrote: "You observed that the hands were anxious to know about working another year. You can state to them that I am willing to give them what they had last year and probably in a few instances I shall be willing to allow some of the children a little more for the ensuing year."[48] All details would be sorted out after his arrival. Two years later he was still offering advice on this subject: "As days are now short and cold and much time is taken up by those people who do not like work very well in thrashing their hands therefore under these circumstances you will discontinue all you can."[49]

Although supervision of the Webster property absorbed his time, Slater preferred this method to the employment of an outside manager who might prove dishonest, inefficient, or lazy. Besides John, two other sons, George Basset and Horatio Nelson, served long apprenticeships under their father.

Slater completely dominated his sons. If they wanted a share in the family business, his sons had to obey his orders, including those pertaining to their education, marriage, and career choices. None

47. Slater MSS, Samuel Slater and Sons, vol. 235, Samuel Slater to John Slater, March 30, 1821; see also White, *Memoir of Samuel Slater*, pp. 241–242.
48. Slater MSS, Samuel Slater and Sons, vol. 235, Samuel Slater to John Slater, February 23, March 4, March 27, and May 4, 1826; March 12, March 15, and November 16, 1828.
49. Ibid., Samuel Slater and Sons, vol. 235, Samuel Slater to John Slater, November 16, 1828.

of the Slater children attended college; all worked for their father. Successful manufacturers, Slater believed, "employed their families in the labors of the business, and, to the extent of this savings of the wages of superintendence and labor, realized the gross profits of manufacture."[50] George, John, and Horatio Nelson Slater did not exercise significant authority in the business until 1829, when Samuel Slater formed a family partnership, Samuel Slater and Sons.

DECLINE OF THE SLATER SYSTEM

Weaknesses within the Slater system became increasingly apparent in the 1820s. The Slater business network included cotton factories, woolen mills, commission firms, foundries, machine shops, and real estate in several states. The business grew in a topsy-turvy manner, and scant attention was directed toward coordinating and integrating the various concerns. The administration of these units presented another set of difficulties. In some cases Samuel Slater tried to exercise personal oversight of operations, especially at Webster and Pawtucket. Unwilling to delegate authority, he made most of the decisions, including those concerning day-to-day operations. This form of administration occasioned delay and proved inefficient and troublesome. When Slater did delegate responsibility, it was to men he had chosen on the basis of friendship or kinship rather than ability.

Several years before his death, Samuel Slater recognized that his values and his way of doing business were out of step with the practices of those around him. The panic of 1829 highlighted these differences. During the 1820s the textile industry had boomed. "THE COTTON MANUFACTURE is increasing at a wonderful rate in the United States," boasted the *Niles Weekly Register* in 1828. "Many of the old mills are worked to their utmost production, and new ones are building or projected, in all parts of our country. The more the better."[51] But this optimism was premature. Many of these new firms were poorly equipped and improperly managed by agents

50. Ibid., Samuel Slater and Sons, vol. 235, Samuel Slater to John Slater, April 22, 1831; *McLane Report*, 1:928.
51. *Niles Weekly Register*, June 28, 1828.

and owners, and during the economic downturn in 1829, some of them went under while others were forced temporarily to suspend operations. Banks closed, money was in short supply, and credit dried up. A spate of bankruptcies ensued.[52] Even the most successful firms, such as those owned by Samuel Slater, were not immune. In January 1829 Slater acknowledged: "It is rather a pinching time here for money; . . . I have a very heavy load on my back, &c. It is true, I am on two neighbours' paper, but am partially secure, and hope in a day or two, to be fully secured against an eventual loss."[53] The expected upturn did not come, however, and by the early summer recovery appeared remote. Slater found himself stretched to the limit. The two neighbors referred to were David Wilkinson and a family friend, John Kennedy. He had endorsed their notes (which totaled about $300,000), and when the two men failed, he fell liable for their debts. To a group of fellow businessmen he wrote: "D. W. [David Wilkinson] has gone down the falls. His failure is a serious one, and it affects my mind and body seriously, and purse too for the present, but hope eventually to meet with but little loss."[54] This hope was in vain. Debts mounted as another transaction went sour. To a former employee, Olney Robinson, Slater had lent between $4,000 and $5,000 to construct a factory; Robinson failed.[55] Unable to raise the funds he required immediately, Slater was forced to issue thirty notes against his vast business network, including claims against his factories. He did not come away from the crisis unscathed. Shares in several factories were sold.[56]

Although these financial losses proved temporary, Slater feared the future. Friendship, kinship, and trust had been the basis for several loans made by Samuel Slater, and when his associates went bankrupt, Slater's confidence in himself as a businessman was shaken. Personal favors for friends almost ruined him. The values that had guided his decisions and policies for decades meant little in the

52. Gilbane, "Social History of Samuel Slater's Pawtucket," p. 492.
53. Samuel Slater to Moses Brown, Cyrus Butler, Brown & Ives, January 7, 1829, quoted in White, *Memoir of Samuel Slater*, p. 246.
54. Samuel Slater to Moses Brown, Cyrus Butler, Brown & Ives, July 29, 1829, quoted in White, *Memoir of Samuel Slater*, p. 247.
55. Conrad, "Evolution of Industrial Capitalism in Rhode Island," p. 325.
56. White, *Memoir of Samuel Slater*, pp. 244–248; "A Financier of the Old School," *Proceedings of the Worcester Society of Antiquity* 5 (1879):9–10; see also *Niles Weekly Register*, June 28, 1828, July 4, 1829.

financial marketplace. Indeed, many of Slater's associates deserted him. When he needed money, for example, he approached William Almy, a partner and an associate for almost four decades, for a loan; but for William Almy, friendship proved insufficient collateral for a loan. In fact, Almy recognized that Slater was vulnerable and seized the opportunity to acquire both the Pawtucket and the Slatersville property. Although later Slater bought back the Slatersville property, his active participation in Pawtucket ended. Slater was out of step with the rest of the business community; competiton had replaced personal relationships and cooperation.[57]

George White noted significant changes in the man after 1829. Unaccustomed to failure, Slater "never before knew what it was to be unable to meet every demand, and could generally anticipate calls. He said to me 'I felt the more, because I had never been used to it.' He felt his dignity as a businessman hurt."[58] The years following the panic caused further alarm, and when business conditions took another downturn in 1834, Samuel Slater considered closing his factories.[59]

Samuel Slater could not adapt easily to a new business morality; nor could he cope with the changes evident in the economy. Between 1790 and 1835 a market economy had emerged. According to Karl Polanyi, "a market economy is an economic system controlled, regulated, and directed by markets alone."[60] In order for the economy to function efficiently in the early nineteenth century, markets had to be developed for all elements of industry, including labor, land, and money. "In a commercial society their supply could be organized in one way only: by being made available for purchase. Hence, they would have to be organized for sale on the market—in other words, as commodities."[61] Labor, for example, would have to be transformed into a commodity and be offered for sale. Market value would dictate the prices paid for this commodity. This trend represented a reversal of traditional relationships between

57. White, *Memoir of Samuel Slater*, pp. 244–245.
58. Ibid., p. 245.
59. Slater MSS, Samuel Slater and Sons, vol. 235, Samuel Slater to John Slater II, January 19, 1834, quoted in Conrad, "Evolution of Industrial Capitalism in Rhode Island," p. 47.
60. Karl Polanyi, *The Great Transformation* (Boston, 1957 [1944]), p. 68.
61. Ibid., p. 75; see also Bendix, *Work and Authority in Industry*, p. 47.

the economy and society. "Instead of economy being embedded in social relations, social relations are embedded in the economic system."[62]

Samuel Slater thought of the world and his relationship with fellow business associates and laborers in different terms from those demanded by the growing market economy. He was no cost-conscious factory master who treated friends and laborers as commodities. Pecuniary interests alone did not dictate his actions. Yet economic survival in Jacksonian America required that costs be calculated and reduced. The near collapse of the business in 1829 suggested that there might be other weaknesses. If Samuel Slater would not institute cost-saving cuts, his sons would. When George, John, and Horatio Nelson Slater gained control of the family business, they directed it along a new path. While their father lived, they proceeded slowly, for Slater relinquished power only reluctantly. Full economic independence for his sons came only with his death in April 1835. At that time his sons ranged in age from twenty-seven to thirty-one. While on vacation in St. Croix, John Slater II died in 1838. Five years later, George B. Slater died of tuberculosis.[63] Horatio Nelson Slater became the architect of a new system. Based on the model established by his father, it nevertheless represented a pragmatic response to contemporary economic problems.

THE LOWELL SYSTEM

The Slater system was reproduced in hundreds of mill towns throughout the northern states. Emulation of the Pawtucket factory began almost immediately. As early as 1813, the *Utica Patriot* carried advertisements for families to join the Slater-style factories

---

62. Polanyi, *Great Transformation*, p. 57; see also Christopher Clark, "The Household Economy, Market Exchange, and the Rise of Capitalism in the Connecticut Valley, 1800–1860," *Journal of Social History* 13 (1979:169–173; Judith McGaw, "The Sources and Impact of Mechanization: The Berkshire County, Massachusetts, Paper Industry, 1801–1885, as a Case Study," Ph.D. dissertation, New York University, 1977, pp. 157–177.

63. Slater MSS, Sutton Manufacturing Company, vol. 1, Company Minutes, 1838; and Webster, Mass., Vital Statistics, Death, 1843.

under construction in the Whitestown region of Oneida County. Less than a decade later, thirteen factories were operating there.[64] Farther south, in the Rockdale district of Pennsylvania, the industry boomed.[65] But these centers could not rival in size and importance the tristate region of southern New England, encompassing parts of Rhode Island, Connecticut, and Massachusetts. It became the center for Slater-style factories. In the Smithfield area of Rhode Island, a sparsely settled region traversed by the Blackstone and Branch rivers, for example, dozens of small Slater-style factories appeared, and industrial colonies grew up around them. By midcentury Smithfield was one of the most highly industrialized regions in New England. The Georgia Cotton Manufacturing Company and its industrial village of Georgiaville were typical of the many new concerns. Constructed in 1813, this stone factory measured 80′ by 36′ and contained 1,000 spindles. About six families worked in the factory and lived in the community that grew up around it.[66] Similar factory communities emerged, including Woonsocket, Central Falls, Lonsdale, Greenville, Manville, and Clayville in Rhode Island; Norton, Mansfield, and Easton in Massachusetts; and Pomfret and Thompson in Connecticut. Whether they were located in Whitestown, Rockdale, or Easton, these firms carried the earmarks of the system first developed by Samuel Slater. Most of the mills were small-scale enterprises in which two or three men pooled their capital and their talents to construct and manage a purpose-built spinning mill. Located in rural areas for the most part, they relied on water as their source of motive power. Furthermore, in most of these factories ownership was synonymous with management and family labor characterized the type of work force recruited.

Although widespread, the Slater system represented only one model for early industrialization. A different system introduced by Francis Cabot Lowell spread throughout northern New England. The Lowell system gained and retained a national reputation in

64. Mary P. Ryan, *Cradle of the Middle Class: The Family in Oneida County, New York, 1790–1865* (Cambridge, Mass., 1981), pp. 43–59.
65. Anthony F. C. Wallace, *Rockdale: The Growth of an American Village in the Early Industrial Revolution* (New York, 1978 [1972]).
66. Steere, *History of the Town of Smithfield*, pp. 104, 94–131.

the industry, and it was considered the legitimate precursor of American big business.

Francis Cabot Lowell began his career not in the factory, like Samuel Slater, but in the countinghouse. As a supercargo aboard his uncle's ship, Lowell learned firsthand the problems associated with the import-export business. This venture did not command his complete attention, however, and he began to speculate in bulk commodities and to invest in manufacturing and real estate. In the years preceding the War of 1812, he made inquiries into textile manufacturing and learned about it initially from his uncles and then on a visit to England, where he observed new weaving machines. On his return to the United States, he described the novel loom to Paul Moody, a local mechanic, who constructed a crude wooden machine for him. Run by water power, this machine produced a sturdy cut of cloth suitable for the growing Western trade. Lowell decided to spin the yarn for the machine himself rather than purchase it from other manufacturers. With eleven associates he formed the Boston Manufacturing Company in 1813 and constructed a factory at Waltham, Massachusetts. Here the two processes of spinning and weaving were carried out under the same roof, a move that altered the American textile industry forever. Still, operations stalled. The adoption of the loom necessitated the development of other equipment, and within a decade, winders, dressers, speeders, and filling frames were introduced, and most of the technological inventions and innovations associated with the Waltham system were set in place.[67]

Innovations continued. Young, unmarried women and girls drawn from rural communities throughout New England entered the factories. To attract and retain this form of labor, Lowell paid cash wages and designed a community to satisfy the needs and demands of this itinerant female labor force. For the women workers, life in Lowell represented a break with the past. Living in company-owned boardinghouses, working from 5:00 A.M. to approximately 5:30 or

67. Jeremy, *Transatlantic Industrial Revolution*, pp. 92–103, 180–203; Ware, *Early New England Cotton Manufacture*, p. 63; Batchelder, *Introduction and Early Progress of the Cotton Manufacture*, pp. 72–73; Appleton, *Introduction of the Power Loom*, pp. 13–14; Robert V. Spalding, "The Boston Mercantile Community and the Promotion of the Textile Industry in New England, 1813–1860," Ph.D. dissertation, Yale University, 1963, pp. 11–13.

7:00 P.M. six days a week, the women in this community lived lives that revolved primarily around the mills. Without the immediate support of a kin network, they came to depend on their own resources and one another. Many developed a sense of independence.[68]

Lowell not only introduced new forms of technology and factory organization and a distinct form of labor, but also diverged from traditional practices in the area of ownership and management. Lowell's first project required about $300,000, more capital than he could raise personally. To solve this dilemma, he applied for and received a charter of incorporation from the Commonwealth of Massachusetts. Shares in the new venture were sold to Boston associates, friends, and relations. A three-tiered system of administration was introduced to manage the new firm. At the top stood the board of directors, comprised largely of major stockholders. They selected a treasurer and a skilled factory agent, the two most important positions in the organization. The first treasurer, Patrick T. Jackson, who was Lowell's brother-in-law, was a stockholder in the firm. Through personal visits and correspondence, he supervised closely the factory agent, a technician who was responsible for operating the factory and overseeing various services connected with it.[69] Although the corporate form of ownership and professional management characterized the administration of this firm, in practical terms the firm continued to be managed like a traditional mercantile concern. The number of initial subscribers was small, and they kept in constant touch with Lowell, Jackson, and the factory agent. Owners actively participated in management decisions.[70]

The factory system introduced at Waltham represented both continuity and change. New ownership and management techniques were introduced, but the circle from which Lowell drew his investors was small, and the Boston Manufacturing Company resembled an extensive partnership network. Ownership was sepa-

68. *Niles Weekly Register*, July 6, 1833; see also Dublin, *Women at Work*, p. 95.
69. Chandler, *Visible Hand*, pp. 68–72; Appleton, *Introduction of the Power Loom*, p. 29.
70. Paul F. McGouldrick, *New England Textiles in the Nineteenth Century* (Cambridge, Mass., 1968), p. 21; Appleton, *Introduction of the Power Loom*, pp. 8, 24.

rated from management, but owners were employed as treasurers and only supervisors were hired from beyond a kinship or friendship network. A unique female labor force was employed, and a degree of paternalism could be observed in labor-management relations.

The experiment at Waltham proved successful. As Nathan Appleton, one of the initial subscribers to the scheme, noted: "From the first starting of the first power loom there was not hesitation or doubt about the success of this manufacture."[71] The *Niles Weekly Register* labeled it "the pride of America."[72] Sales rose from $23,628 in 1816 to more than $260,000 four years later. With the business well established, Lowell began to diversify. He set up a machine shop and offered to sell Waltham equipment to other manufacturers. Patrons, however, had to purchase an entire set of machines consisting of looms, dressing frame, and warping machine. On these sales the corporation's profit was approximately 25 percent. Dividends climbed as the return on invested capital rose to 17 percent in 1817.[73]

With such obvious financial rewards in view, stockholders devised plans to reap the maximum benefit from the business, and after Francis C. Lowell died in 1817, no one restrained them. Charges for equipment doubled; the associates declared a special stock dividend of $50,000 in 1820 and decided to issue new shares of stock to finance expansion rather than to pay for it from existing profits. Furthermore, the directors ordered that the proceeds from this new issue "shall belong to the present company & be divided among them as part of their profits."[74] The Waltham subscribers shared $50,000 in special dividends plus another $90,000 in additional stock sales. In 1822 earnings from machinery together with the subsequent sale of the machine shop itself brought in another $152,000. Of this amount $100,000 was designated as income by the directors and distributed to the shareholders. A continued rise in profits was reflected in the dividends declared: 25 percent in

71. Appleton, *Introduction of the Power Loom*, p. 10.
72. *Niles Weekly Register*, October 5, 1822.
73. Spalding, "Boston Mercantile Community," pp. 28, 22; Ware, *Early New England Cotton Manufacture*, p. 70.
74. Directors Records, Boston Manufacturing Company, February 25, 1820, quoted in Spalding, "Boston Mercantile Community," p. 31.

1823, 25 percent in 1824, and 35 percent in 1825. The Waltham
experiment taught stockholders that the industry offered rewards
beyond the profits derived from the manufacture of cloth. Equipment sales could be lucrative, and considerable profit could be made
from stock manipulation.[75]

Several proprietors decided to build on the success of the parent
company by starting up another concern farther north, in Chelms-
ford, Massachusetts. In 1821 Nathan Appleton, Paul Moody, Patrick Tracy Jackson, and the Boott brothers formed an association,
the Merrimack Company, to purchase land and water rights in
northern Massachusetts and construct a new factory based on the
Waltham model. Capitalized at $600,000, the new firm issued 600
shares of stock to five investors. Each stockholder approached
friends, associates, and kin to take shares. Within four months the
number of shareholders rose to forty-seven, with each investor
subscribing to approximately $12,500 worth of stock.[76] In 1827 the
Merrimack directors voted to establish a second corporation, the
Hamilton Company, but before plans could be concluded, a new
program was advanced. Directors of the Merrimack Company decided to concentrate on manufacturing and to divest themselves of
land, water privileges, and the machine shop recently acquired from
the Boston Manufacturing Company. They resurrected a firm, the
Proprietors of the Locks and Canals on Merrimack River, organized orginally in 1792, and turned assets over to it. The Locks
and Canals Company subsequently became the architect of the new
industrial community called Lowell.[77] The number of investors in
these firms remained small, and they were drawn largely from those
who had participated in the Boston Manufacturing Company. A

---

75. Spalding, a keen critic of the Boston Manufacturing Company, wrote:
The sale of the shop appears to have been a carefully engineered transaction
to drive up the price of Boston Manufacturing stock to enable the promoters
to liquidate their holdings at a profit. James Jackson, Patrick Tracy Jackson
and Israel Thorndike, the largest shareholders, all disposed of the major portion of their shares. After the transfer the value of Boston stock fell steadily,
reaching a low of $600 during the 1829 recession. The company later deval-
uated its stock and its performance was undistinguished. [Spalding, "Boston
Mercantile Community," p. 41]
76. Ibid., pp. 31–36.
77. Appleton, *Introduction of the Power Loom*, p. 28; George S. Gibb, *The Saco-Lowell
Shops: Textile Machinery Building in New England, 1813–1949* (Cambridge, Mass., 1950),
pp. 66–68.

balance sheet for Francis Cabot Lowell II dated December 31, 1829, for example, listed shares in the Merrimack Company, the Hamilton Company, the Locks and Canals Company, the Boston Manufacturing Company, and two small ventures, the Amesbury factory and the Newton mills.[78]

This situation began to change, however, as new factories were constructed in Lowell. In the 1830s Amos and Abbot Lawrence, owners of a large commission business in Boston, entered the village to organize the Tremont Mills, the Suffolk Manufacturing Company, and the Lawrence Manufacturing Company. The three mills organized by the Lawrence brothers were capitalized at between $600,000 and $1.2 million each, and the Lawrence brothers had to cast widely for investment capital. Approximately 128 people subscribed to stock in the Lawrence Manufacturing Company alone, and many of them had little previous experience in the textile industry. By the 1830s manufacturing had become a profitable and secure investment.[79]

Proprietors of these mills and two subsequent ventures, the Boott Cotton Mills (established in 1835) and the Massachusetts Cotton Mills (completed in 1839), had to purchase mill sites, property, equipment, and water privileges from the Locks and Canals Company. Costs for equipment alone ran from $20 to $22 a spindle. Compared to prices paid by manufacturers elsewhere, this was exorbitant. In July 1836 the Amoskeag Manufacturing Company paid $14 a spindle for its machinery, approximately one-third less than the cost of the Lowell equipment. Nevertheless, investors scrambled to secure shares in the new corporations.[80] Amos Lawrence, for example, thought the Boott Cotton Mills a "good concern" and purchased shares for himself and his sons.[81] Lawrence had a right

78. Lowell MSS, Schedule of Property of Francis Cabot Lowell II, Balance Sheet, December 31, 1829.

79. Lance E. Davis, "Stock Ownership in the Early New England Textile Industry," *Business History Review* 32 (1958):208. Davis estimated that mercantile interests represented over one-third of the investment holdings in textile manufacturing. See also Henrietta M. Larson, "A China Trader Turns Investor: A Biographical Chapter in American Business History," *Harvard Business Review* 12 (April 1934):350–357.

80. Lowell MSS, Permanent Investment of the Amoskeag Manufacturing Company, July 1, 1836.

81. Amos Lawrence to William Lawrence, Boston, April 29, 1835, Amos Lawrence Papers, Massachusetts Historical Society, Boston.

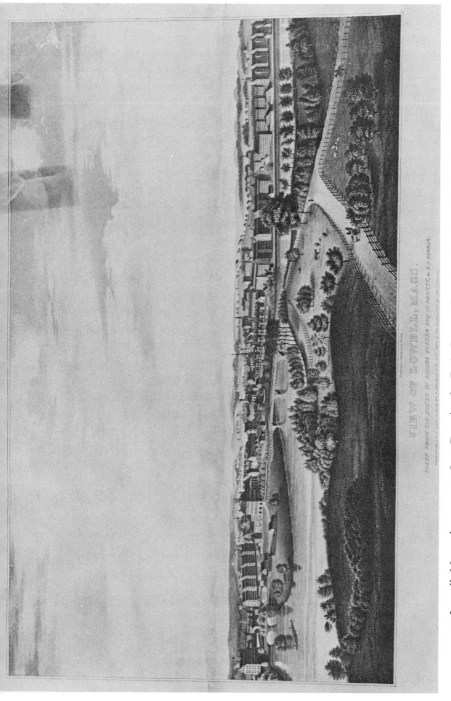

VIEW OF LOWELL, MASS.

3. Lowell, Massachusetts, c. 1834. Drawing by E. A. Farrar. Courtesy Merrimack Valley Textile Museum, North Andover, Massachusetts.

to be optimistic. In 1832 the Appleton, Merrimack, and Hamilton companies manufactured 4,275,849, 6,460,000, and 3,650,000 yards of cloth, respectively. Even during periods of economic dislocation and labor unrest production schedules were maintained. For the six months ending January 1837, John Aiken, factory agent for both the Tremont and Lawrence mills, reported that more than 8 million yards of cloth had been manufactured by the two mills and that the quality "has been good in the main, *very good*."[82] During the period operators had walked out, forcing the factories to suspend operations for several weeks. Still returns on capital at Lowell remained high. At the Suffolk mill profits for the period from 1831 to 1839 averaged over 17 percent annually, while the Merrimack Company averaged 14 percent of net worth between 1828 and 1835.[83] Virtually all earnings were paid out in dividends, and whenever additional capital was required for expansion or renovation, new stock was issued.[84]

With the success of these firms additional factory sites were developed at Taunton, Three Rivers, Nashua, Saco, Canton, and Manchester, and men of the stature of John Cushing, a retired China trader, competed with the Lawrence, Appleton, and Thorndike families for new stock issues.[85] While new investors found this industry attractive, most of them avoided active participation in the management of the concerns. Many investors removed themselves totally from the business and hired such firms as Bryant and Sturgis to act on their behalf. For such men the textile business represented a source of income. Like Amos Lawrence, most of them were "unwilling to bear any loss as a stockholder."[86]

Administration of the new firms fell to a small group of stockholders. The three-tiered system of management first evident at Waltham continued to be employed in Lowell: board of directors, treasurer, factory agent. The position of treasurer was an espe-

82. *McLane Report* 1:340–341; John Aiken to A. Lawrence, February 13, 1837, February 22, 1837, Amos Lawrence Papers.
83. McGouldrick, *New England Textiles in the Nineteenth Century*, p. 84; Balance Sheet, December 19, 1839, Amos Lawrence Papers.
84. McGouldrick, *New England Textiles in the Nineteenth Century*, pp. 121–138.
85. Larson, "China Trader Turns Investor," p. 354; Jackson Company, List of Stockholders, June 23, 1838, Amos Lawrence Papers.
86. Amos Lawrence to A. A. Lawrence, December 4, 1836, Amos Lawrence Papers.

cially sensitive one. Until the 1830s each man who accepted the position defined it in his own terms, and some men devoted more time to it than others. The actual time spent on mill business became an issue, and stockholders wanted to spell out the duties, responsibilites, and remuneration attached to the office. In 1837 Kirk Boott, treasurer of the Merrimack Company, died and was replaced by Francis Cabot Lowell II. When Lowell's name was announced, investors and other interested parties sought to regularize the line of authority between the directors, the treasurer, and the factory agent. Edward Brooks, who was related to the Boott family through marriage, was concerned that Lowell would not devote enough attention to the firm. In a letter to Lowell in 1837, he suggested that "it should be distinctly understood that the whole time of the agent was to be devoted to the Business of the Company to the exclusion of any other occupation," and he expressed concern that Lowell had too many outside commitments.

> In your case, I presume from the conversation we have had together, and from your well known scrupulousness in the performances of whatever you undertake that there will be no difficulty on this score. I have understood that you intend to resign immediately the employment you now hold under the city of Water Commissioner, and for myself I should be perfectly willing to leave it to your own sense of duty to decide how far you were justified in taking upon yourself responsibility of any kind.[87]

Brooks suggested that the duties and the responsibilities be outlined clearly so that all parties would know what to expect. Lowell held the position of treasurer for a short time, and then he resigned.[88]

Brooks and others believed that birthright did not qualify a person for such an important position. Competence, diligence, accountability were more important than family connections in the selection of the treasurer and the other managers. Unlike Slater's sons, the sons of Lowell, Appleton, and Lawrence could not expect to assume high positions and important responsibilities in the industry owned by their fathers.

87. Lowell MSS, Edward Brooks to Francis Cabot Lowell, May 23, 1837.
88. Appleton, *Introduction of the Power Loom*, p. 34.

Despite its early financial success and the acclaim it received, the Lowell system had its share of problems. As ownership became more diverse and as management at all levels passed to hired professional men, there was pressure to emphasize immediate gains, to maximize profits, often at the expense of the introduction of new technology, product diversification, or labor-management relations. Over the years the relationship between workers and management deterioriated. This became evident in the 1830s, when an economic downturn forced some mills to curtail operations. At Lowell management responded to dislocation in the industry by raising boardinghouse charges and lowering wages, first in the winter of 1834 and then again in the fall of 1836.[89] Labor retaliated. Any reduction in wages or increase in rent threatened the economic independence of the female operatives. The women understood that their relationship with the companies rested increasingly on an economic basis, not on a community of interests or on any notion of mutual duties and obligations. When their income, and hence their independence, was threatened, they fought back. Thomas Dublin in *Women at Work* explored the sense of economic independence emerging among the working women of Lowell.

> When women workers spoke of independence, they referred at once to independence from their families and from their employers. An adequate wage made them largely independent of their families back home and also allowed them to save enough out of their monthly earnings to return to their native homes whenever they so desired. But this independence was based on the relatively high level of wages in the mills. The wage cuts threatened to deny women these savings and the economic and social independence they provided, offering instead the prospect of a total dependence on mill work.[90]

In the nineteenth century the Lowell and the Slater systems represented two different approaches to industrialization. By the 1830s economic self-interest and indeed a rising sense of individualism came to characterize Lowell-style firms. Ownership and management were separated. Lowell stockholders viewed their relation-

89. Ware, *Early New England Cotton Manufacture*, pp. 98–100; Dublin, *Women at Work*, p. 86.
90. Dublin, *Women at Work*, p. 95.

ship to the business largely in economic terms. While economic self-interest was evident in the Slater system as well, Slater emphasized close product-family identification, family participation and control, and slow, steady long-term growth.

What prompted Slater to adopt this business scheme? Despite his voluminous business correspondence, his public addresses and publications, and the numerous books and articles that have been published about him, Slater remains an enigmatic figure. He was ambitious, determined to succeed, hard-working, and frugal, but these characteristics explain only partly the man's complex nature. Scholars have attributed many of his business decisions to "entrepreneurial conservatism," to economic dislocations in the industry, or even to a desire to preserve his reputation for producing quality goods.[91] These factors do not adequately explain his actions. Although he was one of the most important businessmen in America and one of the most successful manufacturers of the time (by the 1830s his property, buildings, and equipment in Webster alone were valued at $242,000), throughout his long career Slater was bound by traditional or preindustrial values that influenced his actions and decisions. Traditional notions of authority and responsibility were reflected in his business organization and in his labor-management relations.[92]

Slater monopolized power. He not only made all the strategic choices concerning finance, diversification, marketing, and the introduction of new technology, but also remained involved in day-to-day operations. Personal ownership and management could be effective primarily on a small scale. Personal control of dozens of widely dispersed enterprises proved impossible. While the growth of his business empire necessitated new attitudes toward management, Slater could not break with past practices. He could not delegate authority to outsiders.

In the Slater firms, loyalty, trust, kinship, and friendship ties meant more than merit in the choice of partners and managers. This was

---

91. Chandler, "Samuel Slater, Francis Cabot Lowell, and the Beginnings of the Factory System," p. 22; Coleman, "Rhode Island Cotton Manufacturing," p. 66; Bagnall, *Textile Industries of the United States*, p. 399; Ware, *Early New England Cotton Manufacture*, p. 76.
92. *McLane Report*, 1:576–577.

obvious when Slater sent children, his sons, to supervise other children at the Webster property. Neither Samuel Slater II nor John Slater II had technical or managerial experience, and hence neither could fulfill his obligations.[93] This was a family business meant to be handed down to the next generation. Growth and profits ranked behind other, personal considerations.

Slater's commitment to traditional values could be observed not only in ownership and management policies, but also in the type of labor he employed in his factories and in the industrial colonies he established. Family labor predominated, and the communities he planned and built served to attract and retain this form of labor. Through advertisements placed in the *Massachusetts Spy* (Worcester), in the *Plebeian and Millbury Workingman's Advocate*, and in the *Manufacturers' and Farmers' Journal and Providence and Pawtucket Advertiser*, Slater sought to attract people to his factories. Advertisements assured them that they would be provided for and that they would be given steady, honorable forms of labor if they accepted industrial employment. Many New England householders responded to the advertisements, leaving their farms, their jobs, and their communities in search of a better life in the factory towns. On arrival many of them found a style of life and a set of work patterns similar to those they recently had left behind. Within the context of the factory system, Samuel Slater recreated the traditional New England village.[94]

93. See chap. 8.
94. *Massachusetts Spy* (Worcester), October 30, 1822; *Plebeian and Millbury Workingman's Advocate*, January 25, 1832; *Manufacturers' and Farmers' Journal and Providence and Pawtucket Advertiser*, January 12 and February 20, 1826.

[5]

Industrial Community Life

The Industrial Revolution tore the fabric of traditional Western society. Where the transition to an industrial society was swift and where factory masters disregarded traditional values and institutions, social discord and violence often resulted. Yet it was also possible to preserve much of the preindustrial way of life within the context of the factory system, to ease the transition of workers from an agricultural existence to one based on the wage economy and thereby to minimize tension and conflict.

Samuel Slater tried to effect a peaceful transition between agricultural life and industrial life. To this end, he employed family labor almost exclusively during the first generation of factory operations, and he designed industrial communities that preserved such links with New England's colonial past as the open-field village pattern, single-family dwellings, the church, and the town meeting. For scores of rural New England families, Slater's factory colonies served to ease the tension of moving from one line of work to another and from one community to another. As long as Slater consciously sought to preserve these ties with the past, to maintain traditional institutions and lifestyles, he easily attracted family labor to his factories and maintained harmonious labor relations with householders.

FACTORY VILLAGES

Employing a traditional linear street design and an open-field village pattern, the factory towns of Slatersville and Webster re-

[125]

tained the traditional New England sense and appearance of community. Founded in 1806, Slatersville was located in upstate Rhode Island. Like many of the towns of colonial New England, it was built around a broad road that traversed the town center. The smithy, the grocery and dry goods stores, the church, and the school were on this road. Predictably, the Congregational church stood in the geographic center of the village and was surrounded by a broad common. Toward the outskirts of the village lived more than six hundred textile workers, farm laborers, merchants, and mechanics. Their homes were one- and two-story detached and semidetached dwellings that were built parallel to the main road and separated from one another by garden plots. Each dwelling was occupied by a single family. No house stood isolated from the central community. The mill and its outbuildings, which were owned by Almy, Brown, and the Slaters, did not disturb the traditional sense of community. They were built at a short distance from the village and were surrounded by fenced and tilled fields belonging to the company.[1]

Webster, Massachusetts, founded in 1811 and incorporated in 1832, also preserved the traditional New England town design and sense of community within the context of the factory system. A few miles north of the Rhode Island–Connecticut state border in the South Gore region of Oxford Township, Slater built three mill villages, two of them along the French River and the third, East Village, several miles away, near Chaubunagungamaug Pond. East and North villages grew slowly, and by 1861 they contained fewer than three hundred residents. A mill, a Baptist church, and about twenty-five company dwellings comprised East Village, while North village was smaller yet, containing only a mill, some tenements, and a

1. In the field of nineteenth-century industrial architecture America ranks high. Lowell, Lawrence, and Pullman easily captured the imagination and the attention of the world. Less spectacular in concept and design, the many small industrial towns and villages scattered over New England and the Middle Atlantic states nevertheless represented a unique response to industrialization. For a discussion of mill architecture see Henry Russell Hitchcock, *Rhode Island Architecture* (Cambridge, Mass., 1939), pp. 36–40; Bryant Tolles, "Textile Mill Architecture in East Central New England: An Analysis of Pre-Civil War Design," *Essex Institute Historical Collections* 107 (July 1971):223–253; A. N. Fowler, "Rhode Island Mill Towns," *Pencil Points* 22 (1936):271–282; and Richard M. Candee, "The Early New England Textile Village in Art," *Antiques* 98 (December 1970):910–915. I thank Mr. Candee for sharing his slides of factories and industrial communities with me.

4. Almy, Brown, and the Slaters, Slatersville, c. 1823. Courtesy Slatersville Congregational Church, On-the-Common, Slatersville, Rhode Island.

boardinghouse. South Village became the focus of community life in Webster, and it is this village that is described here.[2]

Like Slatersville, Webster followed the traditional open-field village pattern and its concomitant street design. Along Main Street were located shops, stores, the post office, the tavern, the hotel, the Congregational church and, later, the Methodist church, and the detached and semidetached homes to laborers, operatives, artisans, and merchants. In 1832 approximately 1,170 people lived in the township, and the community continued to grow so that by 1840 the population numbered about 1,400. Beyond the village center stood the mills, and beyond them were over seven thousand acres of arable land, pasture land, and woodlots belonging to the company or to freehold farmers.[3]

In Webster even architecture represented a carry-over and a concession to preindustrial society. Multifamily, block-style tenements and barracks-like boardinghouses, common in Lowell and Chicopee, were slow to appear in Webster. Although at first the company encouraged workers to find housing in the community and in the surrounding area, later it built a small boardinghouse and about twenty-five family dwellings.[4] Located on the edge of the village, these stone cottages conformed in broad outline to traditional concepts and design. The front door opened onto a large kitchen approximately fifteen feet square; beyond the kitchen were two additional rooms, one probably serving as a parlor and the other as a bedroom. Two sets of stairs led off the kitchen, one to

2. La Porte, "Birth of America's Spinning Industry, II," p. 676; "Two Hundred Years of Progress," p. 4; Slater and Sons, *Slater Mills at Webster*, p. 25; Charles Reding, *A Historical Discourse Delivered on the Fiftieth Anniversary of the Organization of the Baptist Church, Webster, Massachusetts, 30 October 1864* (Webster, 1864), p. 7; McLane *Report*, 1:576–577; *Leading Business Men of Webster, Southbridge, Putnam, and Vicinity; Embracing Also Danielsonville, East Douglass, and Oxford* (Boston, 1890), pp. 1–5.

3. *Leading Business Men of Webster, Southbridge, Putnam, and Vicinity*, p. 4; Slater and Sons, *Slater Mills at Webster*, pp. 10–21; F. W. Beers, *Atlas of Worcester County, Massachusetts, from Actual Surveys by and Under the Direction of F. W. Beers, Assisted by George P. Sanford & Others* (New York, 1870), pp. 93–94; Hurd, *History of Worcester County*, 1:373. "Two Hundred Years of Progress," p. 4; United States Federal Population Census, Manuscript Schedules, Sixth Census of the United states, 1840, Webster, Mass. (Unless noted otherwise, all citations to the United States Census are for Webster.)

4. Slater MSS, Union Mills, vol. 195, June 25, 1849; General Box, Samuel Slater and Sons to D. W. Jones, September 26, 1836; Manuscript Schedules, Seventh Census of the United States, 1850.

the cellar below and another to the two bedrooms above. The only concession to economy was the common center wall dividing the residences.[5] Two houses formed a unit.

A rural atmosphere characterized this community. Trees were plentiful. Because land was available, each cottage unit was set back from the road on a plot of ground large enough to allow families to plant vegetable gardens. Other rural traditions survived within this new industrial order. Those who migrated to Webster often brought dairy cows, horses, and cattle with them, and they were allowed to graze the animals on company land. In their cottages or in buildings or coups attached to their dwellings, workers kept pigs, goats, chickens, and other small farm animals. Like the vegetables grown in their gardens, these animals provided part of the food consumed by the factory families.[6]

Throughout southern New England, in Harris, Clayville, Fiskville, Hope, Arkwright, and Manville—all Slater-style communities—similar industrial villages emerged. The Manville factory was constructed along the Blackstone River. Its proprietors purchased a large tract of land around the mill and constructed an industrial village, a community described in glowing terms by one local historian: "The village, which lies on the Smithfield side of the river, is well built on wide streets, shaded with beautiful maple and elm

5. Slater and Sons, *Slater Mills at Webster*, pp. 10–21; Beers, *Atlas of Worcester County*, pp. 93–94; Hurd, *History of Worcester County*, 1:373; Fowler, "Rhode Island Mill Towns," pp. 274–282; Hitchcock, *Rhode Island Architecture*, pp. 39–40. In Webster some of the early Slater cottages have survived. Mrs. S. Galvin kindly allowed me to view hers.

6. Webster, Mass., Town Meeting, Minutes: April 2, 1832, and April 7, 1834; see also Slater MSS, Union Mills, vol. 117, Storrs to H. N. Slater, May 5, 1845; vol. 110, Farm Day Book, 1848–1861; vol. 101, Memo of Cows at Hay, 1837–1849; vol. 169, Caleb Brown to Samuel Slater, December 10, 1829; White, *Memoir of Samuel Slater*, pp. 126, 130; "Regulations in Relation to the Tenements Belonging to the Blackstone Manufacturing Company, 1867," Blackstone Manufacturing Company Collection, Rhode Island Historical Society, Providence. In these industrial communities waste disposal posed a serious problem. Probably two to four residents shared a privy, and no one accepted responsibility for its maintenance and cleanliness. Moreover, at no time did local officials pass an ordinance for the collection or the removal of garbage. Instead, swine served as scavengers. Material not consumed by them was left to rot on the roadways or was dumped into the river or the pond from which local inhabitants drew their water supplies. But in the first years of settlement, these drawbacks could be endured easily because the population was relatively small. See Webster Woman's Club, *Now and Then: A Webster Scrapbook, 1832–1932* (Webster, n.d.), pp. 103–104; Fowler, "Rhode Island Mill Towns," pp. 274–282.

trees. It is perfectly kept and evinces the results of careful oversight."[7] In these factory colonies built during the first decades of
the nineteenth century, the design of cottages approximated closely
those found in Webster except that many of them were detached,
not semidetached, dwellings. Constructed of either timber or stone,
these one- or two-story homes contained five or six large rooms. In
these villages a rural atmosphere was also maintained: cottages were
widely spaced and were surrounded by groves of trees and large
garden plots.[8]

Contrast these villages with Lowell and Chicopee. The barracks-
like housing found there represented a complete break with tradition. Communal living forced the workers to conform to group
norms, which in turn fostered the development of a working-class
ethos. In his excellent article on worker protest in Lowell, Thomas
Dublin stressed:

> To the extent that women could not have completely private lives in
> the boarding houses, they probably had to conform to group norms,
> whether these involved speech, clothing, relations with men, or atti-
> tudes toward the ten-hour day. Group pressure to conform, so im-
> portant to the community of women in early Lowell, played a signifi-
> cant role in the collective response of women in changing conditions
> in the mills.[9]

The designs of Webster and Slatersville, with their emphasis on
single-family dwellings, did nothing to promote such a sentiment.
People turned toward the home for aid, comfort, and encouragement, and not toward their fellow workers. In town planning and
architecture, Webster and Slatersville and scores of other southern
New England factory colonies resembled more the compact rural
colonial villages of Dedham, Andover, and Oxford than the new

7. Steere, *History of the Town of Smithfield*, pp. 104, 94–131.
8. Hitchcock, *Rhode Island Architecture*, pp. 38–40; Fowler, "Rhode Island Mill
Towns," pp. 274–282.
9. Dublin, "Women, Work, and Protest in the Early Lowell Mills: 'The Oppress-
ing Hand of Avarice Would Enslave Us,'" *Labor History* 16 (Winter 1975): 106; see
also R. P. Horwitz, "Architecture and Culture: The Meaning of the Lowell Boarding
House," *American Quarterly* 25 (1973):76–80. For an alternative view of the boardinghouse and its role in the life of each operative see Thomas Bender, *Toward an
Urban View: Ideas and Institutions in Nineteenth-Century America* (Lexington, 1975), pp.
63–75.

5. Mill at Clayville, Rhode Island, c. 1820. Courtesy Rhode Island Historical Society, X3 3104.

bustling industrial towns of Lowell and, later, Chicopee and Lawrence.

FACTORY FARMS

The central residential and commercial districts in Slater's industrial communities were surrounded by agricultural land worked by the company, tenant farmers, and freehold farmers. During the 1830s in Webster Township there were fifty-eight freehold farmers, four factory farms, and a number of homesteads owned by the Slater family and operated by tenant farmers. Of the seven thousand acres of land in the township, Samuel Slater and his sons owned approximately two thousand acres.[10]

Slater subdivided his factory farms—Farm A, Farm B, North Village farm, and East Village farm—into three fields, one for tillage, a second for pasture, and a third left as a woodlot. On arable land he planted a variety of vegetable crops, including beans, carrots, turnips, beets, pumpkins, and potatoes. Cranberries were the only fruit crop grown. Most of the fruits and vegetables grown on the farms were consumed by local factory families or by other townspeople. In addition to these crops, Slater planted oats, corn, and hay, probably to provide winter fodder for the horses, oxen, cattle, swine, and dairy cows kept on the farms. A large tract of land was set aside for the pasture of both factory animals and those belonging to local residents. The woodlot served as a source of fuel, fenceposts, and lumber for Slater and his workers.[11]

Each farm appeared well stocked with machinery and draft animals. (Table 3 shows an inventory of Farm A.) On the farms, the firm operated gristmills and ground corn and rye for hands. Slater also owned a blacksmith shop and a slaughterhouse. The laborers

10. "A Financier of the Old School," *Proceedings of the Worcester Society of Antiquity* 5 (1879): 9–10.
11. Slater MSS, Union Mills, vol. 101, April 11, 1842; vol. 102, memorandum of Farmer's Labor, August 1849. Before 1829 apparently Slater sometimes paid local farmers in the area to pasture animals for laborers. See Slater MSS, Union Mills, vol. 169, C. Brown to Samuel Slater, December 10, 1829; see also *Leading Business Men of Webster, Southbridge, Putnam, and Vicinity*, pp. 4–5.

Table 3. Inventory of stock, Farm A, April 1, 1842

Item	Cost
2 red carts	$4.50 each
2 harrows	9.50
1 break-up plow	4.50
1 break-up plow	5.50
1 new plow	8.50
4 old or broken plows	7.50
2 hay rucks	6.00
1 cultivator	4.00
2 sleds, 400 × 500	4.00
2 ox wagons	70.00
1 yoke oxen, 9 years	88.50
1 yoke oxen, 9 years	70.00
1 yoke oxen, 5 years	77.50
1 yoke oxen, 3 years	55.00
1 yoke oxen, 3 years	55.00
1 yoke oxen, 3 years	48.33
1 yoke oxen, 2 years	32.33
1 Red Stag	50.00
1 winnow mill	9.00
14 tons meadow hay	70.00
14¾ tons good hay	177.00

Source: Slater MSS, Union Mills, vol. 101, Account of Stock on Farm, April 1, 1842.

employed on the farms and in the various shops and mills were householders who lived in the factory village.[12]

Although many householders worked on one of the company farms, those with some capital chose to lease tenant farms. The company owned three or four homesteads that it rented annually to factory families. For use of the land, laborers promised to pay half the taxes on the property, provide half the seed, furnish their own wood, fertilize well all the land planted, and pay a fixed rent.[13] Because of the initial financial outlay, however, few householders were able to take advantage of this opportunity, especially before 1850.

The central village and the company-owned farms and homesteads formed two parts of industrial community life in Webster. The factories formed a third unit. A short distance from the village

12. Slater MSS, Union Mills, vol. 110, January 12, 1850; S. & J. Slater, vol. 15, John Slater to W. A. Blackman, March 13, 1835; Sutton Manufacturing Company, vol. 5, Journal A.
13. Ibid., vol. 101, 1837–1849, especially Lease with Augustus Emerson, April 27, 1837; vol. 87, Lease with Lewis Shumway, April 1, 1849.

stood the Slater mills. In Webster a separation between industrial and community life was maintained.[14] For many unskilled and semiskilled householders, migration to Webster, Slatersville, or other Slater-style communities did not necessitate a complete break with the past. Within the context of the factory system, Slater cushioned the transition from an agricultural way of life to one dominated by the factory system.

THE CHURCH

The link with colonial New England would have been incomplete without the strong presence of religious institutions. In Slater's industrial communities the church came to play an important role in the lives of the laborers and in the life of the community. People saw themselves joined to others first through the family and second through the church.

In Slatersville the Congregational church predominated. Both John Slater, resident manager of the mills, and his wife, Ruth Slater, were members of the church. A devout Congregationalist, Ruth Slater participated in all church affairs, from worship service and prayer meetings to the organization and operation of the local Sunday school.[15]

In Webster the situation was somewhat different. Before mid-century three evangelical denominations established churches in the community: the Baptists in 1814, the Methodists in 1823, and the Congregationalists in 1838. Of the three sects, the Methodist

14. In South Village the woolen mills were built. By 1832 they employed some 128 people. Within the township limits Slater operated several other factories, including Union Mills and the Phoenix Thread Mill, which employed 137 and 43 people, respectively, in 1832. See *McLane Report*, 1:576–577.

15. E. A. Buck, *An Historical Discourse Delivered at the Semi-Centennial Anniversary of the Slatersville Congregational Church, September 9, 1866, and a Tribute to the Memory of Madam Ruth Slater, Who Died June 4, 1867* (Woonsocket, 1867), pp. 21–22; see also Albert Donnell, *An Historical Address Delivered at the Centennial Celebration of the Congregational Sunday School at Slatersville, Rhode Island, September 13, 1908* (Woonsocket, 1908), pp. 16–17, 19–20.

[134]

congregation was the largest and the most influential.[16] Throughout the antebellum period, Methodism flourished in Webster. The
revivals that swept southern Worcester County in the 1820s and
continued with added force throughout the 1830s attracted a large
number of Webster's residents. The first Methodist class meeting
was organized in 1823 by an overseer and a factory hand at the
Slater and Kimball works. Within a decade well over half of the
local residents had become church members. The original meet-
inghouse, built in 1828, had to be enlarged in 1834–35 to accommodate all the parishioners.[17] The Quarterly Meeting Conference
of Webster station announced proudly in 1848 that "the Church is
now in a state of prosperity and union. It is true all are not engaged and zealous, but as a whole there is much devotion to God.
Over one hundred I judge have professed conversion during the
past year, fifty-six of these have joined the church on trial. How
many will be found at Gods right hand eternity alone will reveal."[18]

The church monopolized the leisure of local residents. In addition to Sabbath services, the Methodist church encouraged parishioners to attend prayer meetings, monthly Bible study groups, and
the revivals, love feasts, and camp meetings that occurred throughout the year. Children were expected to participate in the Sabbath
school, and women joined the Webster Relief Society, a Methodist-
sponsored organization whose members made clothing for distribution to the poor.[19]

The Congregational and Baptist churches sponsored a variety

16. "Two Hundred Years of Progress," pp. 35, 37. In Webster, Methodism predominated and formed the foundation for a work ethic. Paul Faler's excellent works
on Lynn, Massachusetts, including "Cultural Aspects of the Industrial Revolution:
Lynn, Massachusetts, Shoemakers and Industrial Morality, 1826–1860," *Labor History* 15 (Summer 1974):367–394, and "Workingmen, Mechanics, and Social Change,"
discuss the connection between the Methodist church and the evolution of a work
ethic. Faler confirms the theory that Methodism was the religion of both workers
and manufacturers.
17. *Historical Sketch and Directory of the Methodist Episcopal Church, Webster, Massachusetts* (Worcester, 1904), pp. 4–5, 7.
18. "Methodist Episcopal Church at the Four Corners," December 21, 1884, and
Records of the Quarterly Meeting Conference of Webster, April 5, 1848, United
Church of Christ, Webster, Mass. (hereafter cited as United Church of Christ MSS);
see also Hurd, *History of Worcester County* 1:367; Reding, *Historical Discourse*, pp. 15,
21; and Slater MSS, H. N. Slater, vol. 33, A. Hodges to H. N. Slater, September 12,
1839.
19. United Church of Christ MSS, Webster Relief Society, 1848.

of meetings, organizations, and associations for their parishioners. The Ladies' Circle of Industry, whose purpose was to raise funds for various charitable events, groups, or associations, was especially popular among women in the Congregational church.[20] Altogether these three churches exercised tremendous influence over local residents. Religious activities absorbed the villagers' time, and the mores and discipline advanced by the various churches guided their conduct and influenced relationships between husband and wife, parents and children, friends and neighbors, proprietors and customers, and employer and employee.

Town Meeting

Long a tradition in New England, the town meeting became an integral part of industrial community life in Slater's towns; manufacturer, merchant, farmer, skilled and unskilled laborer alike participated in local government. Initially those elected to public office included Webster's most prominent citizens: John Day, John Larned, and Benjamin Wakefield, all farmers, and George and John Slater, sons of Samuel Slater. For eight consecutive years George Slater served as a selectman while his brother John represented Webster in the House of Representatives. Within a decade, however, elected officials came to represent a wider cross section of the community as many semiskilled and unskilled laborers began to serve on local government bodies. Charles Tucker, William Andrews, Nathaniel Hunt, and Phineas Houghton, a blacksmith, overseer, day laborer, and farm worker, respectively, served as selectmen in the 1830s and 1840s. Unskilled and semiskilled workers were more prominent in other local positions. Most of the hog reevers, field drivers, pound keepers, and fence viewers were farm laborers and day workers, while a mule spinner served as constable and tax collector, and the superintendent of one of the mills served as town clerk.[21]

20. United Church of Christ MSS, Ladies Circle of Industry, Book 1, April 1844–December 1858, Congregational Church Records. (The Methodist and the Congregational church records are both located at the United Church of Christ, Webster, Mass.)'

21. Webster, Mass., Town Meetings, Minutes: November 1832, April 7, 1834, and March 2, 1835; "List of Selectmen," Webster Town Hall; "Two Hundred Years of Progress," p. 4; see also Slater MSS, Slater and Kimball, vol. 3, 1827–1840.

Residents from all walks of life took a lively and active interest in local affairs.

For over a decade town meetings were conducted with a minimum of fuss or discord as residents relied on customary methods to handle major community problems, including the collection of taxes, the control of animals, and, most important, the relief of the poor. The poor laws enacted in early Webster represented yet another carry-over from the past and emphasized the impact of tradition on problem solving.

During the first half of the nineteenth century residents continued to place the orphaned and abandoned youth, the sick, the elderly, and the indigent in the homes of local families for protection, care, and education. The town paid an allowance to those who accepted such responsibilities. In 1839–40 Phillips Brown, for example, received $75 for the care of Ezekiel Davis, "insane" since birth; Royal Marsh, a local farmer, received $62 for the care of Timothy Wakefield and his wife.[22]

In Webster able adults were often contracted in lots, usually to one of the local farmers. Here the auction or sales system probably operated. This practice had been common throughout colonial New England and could still be found in several other nineteenth-century communities. After the town meeting local residents gathered at the tavern and auctioned off paupers either individually or in lots. As late as 1840 the Webster town records report that two lots of paupers were given over to two farmers.[23]

22. It was not until 1847 that Webster's residents abandoned traditional methods of poor relief. See Webster, Mass., Town Meetings, Minutes: March 11, 1839; March 2, 1840; March 7, 1842; March 1, 1847; March 3, 1851.

23. Descriptions of the auction system elsewhere illuminate the mechanics of the system in Webster. In Andover the auction system was calculated "to punish Sloth and indolence . . . correct vice and immorality establish industry, teach economy and prudence encourage virtue and morality and establish at the same time the means of support on the most Just and economical principles So that he who will not work may not eat" (Andover, Mass., Town Records: March 13, 1821, and March 13, 1827, quoted in Benjamin J. Klebaner, "Pauper Auctions: The 'New England Method' of Public Poor Relief," *Essex Institute* 150 [July 1955]:196–202). See also Steere, *History of the Town of Smithfield*, p. 51; Charles F. Adams, *Three Episodes of Massachusetts History: The Settlement of Boston Bay, The Antinomian Controversy, a Study of Church and Town Government*, 2 vols. (Boston, 1892), 2:729; Daniels, *History of the Town of Oxford*, pp. 222–223; Joseph Merrill, *History of Amesbury, Including the First Seventeen Years of Salisbury to the Separation in 1654 and Merrimac from Its Incorporation in 1876* (Haverill, Mass., 1880), pp. 236, 312.

These methods contrasted sharply with those adopted elsewhere. During the 1820s and 1830s, in communities large and small, the almshouse, the asylum, the workhouse, and the pauper farm had begun to replace traditional family care. For many, the poor had become criminals, and as a punishment they were isolated from the rest of the community, confined to institutions, and put to hard labor. But this was not the case in Webster. For a generation or more the family remained the custodian of the poor. It was not until 1847 that Webster's residents abandoned traditional methods and bought a pauper farm.[24]

In outward form, Webster and many other industrial communities conformed broadly to their colonial ancestors. The force of tradition evident there served to ease the transition of hundreds of workers from rural to industrial life.[25]

24. Webster, Mass., Town Meetings, Minutes: March 1, 1847; March 3, 1851. See also Faler, "Cultural Aspects of the Industrial Revolution," p. 377, Emerson, *History of the Town of Douglas*, p. 55; David J. Rothman, *Discovery of the Asylum: Social Order and Disorder in the New Republic* (Boston, 1971), pp. xiii–xiv; Gerald Grob, *The State and the Mentally Ill: A History of Worcester State Hospital in Massachusetts, 1830–1920* (Chapel Hill, 1966), pp. 16–17.

25. Although the following poem described Pomfret Village, an industrial community founded by Slater's brother-in-law, the sentiments it expresses could easily have been echoed by a Webster or Slatersville resident.

> Sweet vale of Pomfret! 'tis to thee
> I offer now my poesy.
> The spot which boyhood hallowed dear
> Has lived in memory many a year,
> And still within the minstrel's breast
> It gives to his dark hours a zest. . . .
> Thy waterfall, thy busy mill,
> The little school-house on the hill,
> The village bursting into view,
> With its old inn beneath the yew—. . .
> How sweet the thoughts that thou recall,
> Thou never-ceasing waterfall!
> Though years have rolled themselves away
> Since last I heard thy waters play, . . .
> Still when the woes of life assail
> I sigh for thee, sweet Pomfret vale.

See Ellen D. Larned, *History of Windham County, Connecticut*, 2 vols. (Worcester, 1880), 2:548.

[6]

The Family System of Labor

To attract workers to his factories, Samuel Slater tried to construct a bridge to the past. The design of his factory colonies, the architecture of his company dwellings, and the institutions he established in the villages conformed in broad outline to those found throughout much of New England a century earlier. This link with tradition was not cosmetic, and it reached into the workplace. The occupations provided for men, women, and children, the conditions under which they labored, and the settlement of wages conformed to custom. Within the context of the new industrial order, familial values were preserved; alterations in the new economic orientation and structure of a society do not inevitably lead to major changes in its traditional units or beliefs.

DIVISION OF LABOR

In Slater's factory communities, a traditional division of labor based on age, gender, and the marital status of family members emerged. Married men performed customary tasks associated with a rural way of life, such as farming and casual labor, while their wives remained at home to care for the family. Primarily children, adolescents, and unmarried women entered the factory to operate the simple equipment.

Married men refused to enter the mills to tend machines. The factory and its discipline frightened and alienated householders, who compared the factory to a workhouse where people were thrown together, assigned specific tasks, and forced to labor under strict supervision.[1] These attitudes could not be changed, and in order to obtain and retain family labor in his communities, Slater had to offer householders a socially acceptable form of task-oriented employment. In nineteenth-century America this meant agricultural labor. Slater divided his two thousand acres of land in the Webster region into company farms and hired local householders to ditch, set posts, plow, harvest vegetable and cereal crops, or chop and draw wood. During the harvest season women sometimes joined their husbands in the field to bring in the crops. The company provided additional employment opportunities for householders by running herds of animals on land too poor for cultivation.[2]

While the men and, infrequently, the women were employed in agriculture, their children usually worked in the factories. Stephen Moore, for example, worked at North Farm with his eldest son, Phineas, while two of his daughters worked at Union Mills. Two of Moore's best hands on the farm were Orrin Raymond, aged forty-five, and Leighton Brown, aged forty-seven. Three of Raymond's seven children and four of Brown's five children worked for the Slater family in and around the mills.[3]

A drawback of agricultural labor is that it is seasonal. From April to November men worked steadily, but during the winter months only a skeleton crew was required to repair the equipment and prepare for the next season. Tenant farmers, too, had time on their hands.[4] Householders were permitted to shift for themselves, and this caused some alarm among factory masters, such as one Troy, New York, manufacturer who argued:

1. Seth Luther, *An Address to the Working-Men of New England, on the State of Education, and on the Condition of the Producing Classes in Europe and America* (Boston, 1832), p. 35; Stanley Lebergott, *Manpower in Economic Growth: The American Record since 1800* (New York, 1964), p. 125.

2. Slater MSS, Union Mills, vol. 110, Farm Day Book, 1848–1861; vol. 101, 1837–1849; vol. 102, 1840–1851.

3. Ibid., vol. 30, Petty Ledger, 1840–1843; vol. 103, Labour on North Farm, 1843–1846; see also Manuscript Schedules, Seventh Census of the United States, 1850.

4. Slater MSS, vol. 103, Labour on North Farm, 1843–1846. Most hands were paid by the day, although some were paid by the month and others received a fixed rate plus room and board. Whether paid by the day, by the month, or by another scheme, farm hands received wages only for time actually worked.

There are many whose families work in the factories, when the man takes a piece of land on shares, and raises corn and potatoes; but this is a more common practice in the New England states, than with us. When the man cannot be employed to advantage, this may do well, but the leisure hours such a one would have, would be a bad example for the factory hands, and I would prefer giving constant employment at some sacrifice, to having a man of the village seen in the streets or shops on a rainy day at leisure.[5]

Such considerations did not prompt Slater to alter his employment scheme. Besides, Slater's householders seldom remained idle through the winter months. The host of odd jobs available in the industrial community for adult males included hauling goods, painting, making boxes, serving as handyman in and around the mills, fashioning boots and shoes, and weaving cloth.[6] Many men probably turned to hand-loom weaving to supplement the family income. In the United States weaving was performed by both males and females. In the Slater factory communities this type of labor was readily available, especially before 1830. Hundreds of people, employed either by subcontractors or by Samuel Slater, wove the yarn into cloth. The names of farm hands and unskilled workers were prominent on the Slater and Tiffany account books. Although some men undoubtedly collected wages earned by their wives and daughters, others probably operated the looms themselves. From December through February they collected yarn, and they returned finished cloth in April and May. From late spring through December they were free to engage in agricultural work.[7]

Although most male householders remained outside the confines of the factory, some men could be found working in the various departments. All of the skilled and supervisory positions, such as mule spinner, dyer, dresser, machinist, and overseer and overseer's assistant (known as second hand), were filled by men. They comprised 9 to 11 percent of the factory labor force.[8] The firm

5. Letter, Jedediah Tracy, December 26, 1827, quoted in White, *Memoir of Samuel Slater*, p. 130.
6. Slater MSS, Union Mills, vol. 30, Petty Ledger, 1840–1843, especially George Slater, March 1840–April 1843.
7. Ibid., Slater and Tiffany, vol. 80, Hand Loom Weavers, 1812–1829.
8. Ibid., Union Mills, vol. 155, Time Book, 1840.

also employed carpenters and masons to build and repair equipment, to see that the factory was maintained in good order, and to construct fences and perform other jobs in and around the mill. Altogether these men comprised another 14 to 16 percent of the entire factory force.[9] At no time, however, did Slater employ householders to tend the simple machines. Men were not forced to compete directly with their children for work or for wages.

In America more than in England or Europe, men and women remained within clearly delineated spheres where tradition, not necessity, defined the duties and roles of husband and father, wife and mother. Alexis de Tocqueville best described this gender typing so prevalent in America when he wrote that men and women

> are required to keep in step, but along paths that are never the same. You will never find American women in charge of the external relations of the family, managing a business, or interfering in politics; but they are also never obliged to undertake rough laborer's work or any task requiring hard physical exertion. No family is so poor that it makes an exception.[10]

Although American society recognized that many married women had to work, it tried to delineate sharply the form female employment should take. Editors for the *Niles Weekly Register*, for example, advised married women: "We think that relief is to be obtained only in the establishment of such manufacturers as they can carry on at their own homes," for it was in the home that "small children, always requiring the mother's care, may be made useful, and earn, perhaps, their own subsistence."[11]

In Webster most married women followed such admonitions and worked inside or around the home; they cared for the family, performed countless household chores, tended the family garden plots, and, often, took in boarders. When piecework was available, they fashioned buttons, gloves, or palm-leaf hats for local contractors, or they wove cloth for the Slater company or for a merchant weaver. Because of their proximity to the factory, they often found weaving

9. *McLane Report*, 1:576–577; Slater MSS, Union Mills, vol. 155, Time Book, 1840.
10. Tocqueville, *Democracy in America*, p. 601.
11. *Niles Weekly Register*, October 25, 1834.

the most convenient and perhaps, at least initially, the most remunerative occupation to pursue. It also allowed them to supplement the family income while remaining at home. Married women persisted in this occupation despite declining wages in the mid- to late 1820s, when technological improvements in the factory, especially the introduction of the power loom, forced down piece rates. Women who earned $0.06 a yard working for Slater in 1818 received from $0.04¾ to $0.05¾ in 1823 and from $0.02⅛ to $0.03⅜ a yard two years later. By 1825 women worked longer hours and for less money than previously. In 1823 Pearley Foster, for example, earned approximately $55 for weaving by hand 950 yards of gingham, stripe, and shirting. Although two years later she increased her output to 1,050 yards of cloth, she received only $31 for her work.[12] Yet Foster and others continued to weave. Alternative jobs for women were few and generally paid as poorly as weaving. Women entered the straw-bonnet industry, a seasonal occupation and one subject to fashion demand, at one of its many stages from the cutting, smoothing, bleaching, and braiding of the straw to the assembly and trimming of the bonnet. For their labor and that of their children, they received from $0.25 to $0.30 a day, wages unlikely to induce women to leave hand-loom weaving.[13]

While married women and widows participated fully in domestic industry, they were seldom found in the factories working alongside their children. In Slater's industrial villages parents maintained their traditional economic and social positions within the family and the larger community. The jobs they performed and the work patterns they followed conformed to customary practices. Adult men and women had little or no practical experience with the monotony of factory work or the restrictions imposed by industrial discipline. Subordination to the demands of machinery fell on the more vulnerable groups in society as parents, uncles, and brothers who refused to be regulated by the new system themselves willingly placed youngsters in the factory. Child labor was vital to

12. Slater MSS, Slater and Tiffany, vol. 82, Hand Loom Weavers, Pearley Foster, April 1822–May 1826.

13. *Niles Weekly Register*, June 18, 1831; May 17 and October 25, 1834; Gallatin, "Manufactures," October 1, 1809, in *American State Papers Finance*, 2:439; *Manufacturers' and Farmers' Journal and Providence and Pawtucket Advertiser*, January 6, 1820.

the survival of the family, and few unskilled householders could afford to keep their children at home or allow them to attend school throughout the year.

COMPOSITION OF THE FACTORY LABOR FORCE

In the first decades of the nineteenth century, young children, adolescent boys and girls, and unmarried women comprised approximately three-fourths of the industrial labor force. From the outset few people opposed the employment of young people: quite the contrary, society condoned and encouraged it. H. Humphrey, noted author of child-training books, expressed the prevailing attitude toward the employment of children when he wrote:

> Our children must have employment—must be brought up in habits of industry. It is sinful, it is cruel to neglect this essential branch of their education. Make all the use you can of persuasion and example, and when these fail interpose your authority. . . . If he will not study, put him on to a farm, or send him into the shop, or in some other way provide regular employment for him.[14]

And Mathew Carey, a well-known political economist, praised manufacturing as an excellent form of youthful employment: "The rise of manufacturing establishments throughout the United States, elevated thousands of the young people of both sexes, but principally the females, belonging to the families of the cultivators of the soil in their vicinity, and from a state of penury and idleness to competence and industry."[15] Usefulness was still the yardstick by which society measured an individual's worth, and factory work allowed otherwise unemployed or underemployed people to make profitable use of their time and to benefit the community.

The young labor force of Union Mills in Webster was typical. In 1840 thirty children and adolescents worked in the carding department under the direction of an overseer and several second

14. H. Humphrey, *Domestic Education* (Amherst, 1840), p. 132.
15. Mathew Carey, *Essays on Political Economy or the Most Certain Means of Promoting the Wealth, Power, Resources and Happiness of States Applied Particularly to the United States* (Philadelphia, 1822), p. 458.

hands. Approximately two-thirds of the workers there were female; 52 percent were children from nine to twelve years of age, 31 percent were from thirteen to fifteen years, and the remainder were sixteen or older. In the spinning department, the gender ratio approximated that of the carding room. Children as young as eight were introduced to the factory system through employment in this department. Of the twenty-five laborers employed there, 32 percent were from eight to twelve years of age, 44 percent were from thirteen to fifteen, and the remainder were sixteen or older. An overseer and a second hand monitored the labor of the spinning-department employees.[16]

One of the largest rooms in the factory was the weaving department, and there young women, not children, dominated the labor force. At Union Mills in 1840 sixty-nine women wove either full or part time. With the exception of two young sisters, Mary and Sophia Strether, aged eleven and twelve, all of the women employed in the weaving department were between the ages of fourteen and twenty-four. Although some hand-loom weavers remained on the payroll, most of the cloth produced by Samuel Slater and Sons in 1840 was woven by machine.[17]

Most of the people employed at Union Mills belonged to kinship groups. During the eary years of industrialization family labor dominated the factory floors. Slater employed only a few people who had no kin working for a Slater enterprise or who did not live in the factory colonies; such employees were men who assumed skilled and supervisory positions and girls and unmarried women who tended power looms.

The divisions of the labor force between family labor and itinerant hands and other workers was clear in the Slater and Kimball factory in the 1820s and 1830s. In 1827, for example, approximately 102 people worked in the factory: 85 people, or approximately 83 percent of the labor force, were drawn from kinship groups; approximately 9 percent were skilled and supervisory men; and 8 percent were single itinerant women. Conversion to the power loom that year probably caused a temporary dislocation in the labor market. Slater may not have been able to secure immediately

16. Slater MSS, Union Mills, vol. 155, Time Book, 1840.
17. Ibid.

enough women capable of operating these machines from among local factory families, and he searched among farm families in the Oxford vicinity for workers. Six of the eight young women hired to operate the looms that year were Pracy and Joanne Phipps, Maria and Betsey White, Savinia Robbins, and Naomi Morse; all of them had kin living on farms in the Oxford Township area.[18]

The entry of single itinerant women into the factory proceeded haphazardly. In 1832 a number of single women again found work at Slater and Kimball. Ninety-eight of the 132 people employed there that year, or approximately 74 percent, were drawn from kinship groups working in Webster; another 11 percent were skilled and supervisory personnel; and 15 percent were young women hired to operate power looms.[19] (See Table 4.)

While in 1827 and again in 1832 Slater hired a number of itinerant female hands, the practice appeared to be a short-term solution to satisfy an immediate but limited demand for labor. Many women were let go after only one year and were replaced by family help. Furthermore, when such women did not board with their own families in the area, they found accommodation with local factory families and were subject to their government and discipline. While the participation of these workers had implications for the

Table 4. Composition of Slater and Kimball work force, 1827–1835

Year	Families	Single Men	Single Women
1827	25	9	8
1828	23	10	0
1829	21	7	0
1830	22	12	1
1831	23	10	0
1832	29	14	20
1833	26	13	5
1834	9	6	0
1835	8	14	2

Source: Slater MSS, Slater and Kimball, vol. 3, Agreements with Help, 1827–1835.

18. Slater MSS, Slater and Kimball, vol. 3, Agreements with Help, 1827; see also Daniels, *History of the Town of Oxford*, pp. 624, 644, 664–666; Manuscript Schedules, Sixth Census of the United States, 1840.

19. Slater MSS, Slater and Kimball, vol. 3, Agreements with Help, 1832.

future composition of the labor force, in the 1830s their numbers were small, and family labor continued to dominate the work force.[20]

PARENTAL AUTHORITY AND FACTORY OPERATIONS

Under the family system of labor, householders exercised considerable power within the factory, influencing the composition of the labor force, the allocation of jobs in the various departments, the supervision of hands, and the payment of wages. Bargaining between labor and management over the employment of children and labor conditions began before the youngsters entered the mills. On behalf of their children, householders negotiated a contract with Samuel Slater. Casual and verbal compacts at first, these agreements became more formal over time. Although written contracts certainly were initiated earlier, the first set of formal agreements found in the Slater company records are dated 1827; the last are dated 1840. Drawn up in February and March and effective from April 1, the annual contracts made between householders and Samuel Slater listed the names of kin employed, their rate of pay, and any special conditions pertaining to their employment. Typical of these agreements was one signed by John McCausland in 1828:

> Agreed with John McCausland for himself & family to work one year from Apr. 1st next as follows viz:—
> Self at watching 5/6 pr night = provided that any contract made with Saml. Slater for the year shall be binding in preference to this—
> Self to make sizing at 9/- pr. week
> Daughter Jane   —12 pr week—
> Son Alex.   —7/ " "
>   " James   —5/ " "
> each of the children to have the privilege of 3 months Schooling and Alex to be let to the mule spinners if wished.[21]

Education and training provisions were commonly included in the contracts. Parents sought release time from factory employment so

20. The decline of the family system of labor is discussed in chaps. 8 and 9.
21. Slater MSS, Slater and Kimball, vol. 3, Agreements with Help, John McCausland, 1828.

that their children could attend school from two to four months annually, and permission was granted for both boys and girls to attend class. For their sons, householders sometimes sought further concessions. Like John McCausland, many parents wanted their sons to learn a skilled trade such as mule spinning, an occupation that commanded both prestige and high wages.[22]

To ensure that he would have enough operatives to maintain full production schedules, Slater occasionally stipulated that householders supply him with a certain number of laborers or that parents replace those operatives withdrawn from the factories during the year. In March 1829 Abel Dudley, for example, agreed to tend the picker and to place two of his daughters, Mary and Caroline, in the mills. The contract stipulated further that "Mary & Caroline have the privilege of going to School two months each—one at a time and Amos is to work at 4/ pr. week when they are out."[23] Earlier, Thomas Twiss placed three of his children, Thomas Jr., Louisa, and William, in the factory. He agreed that daughter Mary would receive 3 s. per week "to commence when she may be wanted."[24]

While these contracts limited labor turnover and guaranteed Slater a steady supply of workers, they also ensured that parents would retain their position as head of the kinship unit and that children would not gain economic independence. Children looked to their parents to protect their interests. All children employed in the factory had to be sponsored by a householder; with few exceptions before 1830, Slater did not look beyond the kinship unit for labor.

Parents apparently also determined in which department their children would work and the conditions under which they would do so, although this is not stated specifically in the contracts. Family members often worked in the same department, attending machines side by side. Mule spinners hired and paid their sons, nephews, or close family friends to piece for them, and weavers hired

22. Ibid., 1827–1840; see also Pomfret Manufacturing Company Records, Connecticut State Library, Hartford, Contract and Memorandum Books 1 and 2, cited in *The New England Mill Village, 1790–1860*, ed. Gary Kulik, Roger Park, and Theodore Z. Penn (Cambridge, Mass., 1982), pp. 437–461. Smith Wilkinson's contracts began as early as 1807.

23. Slater MSS, Slater and Kimball, vol. 3, Agreements with Help, Abel Dudley, March 19, 1829.

24. Ibid., Thomas Twiss, March 8, 1828.

kin to assist them at their machines. In 1840 Asa Day, a blacksmith employed by Union Mills, Webster, placed his daughters, Francis and Caroline, aged nine and thirteen, in the carding room. John Costis's six children, who ranged in age from nine to eighteen, also worked there, while the four Drake youngsters, aged ten to sixteen, worked together in the spinning department. In the weaving room, sisters often tended looms near one another. Mary Strether worked beside her older sister, Sophia; the Boster sisters, the three Faulkner girls, and the Foster and the Fitts sisters also worked there.[25]

Parental concern did not end with the formal agreement. Although Samuel Slater established strict rules and regulations for the smooth, efficient operation of the factory, and although he demanded that workers be punctual, regular in attendance, industrious, and disciplined, he nevertheless bowed to parental pressures and allowed householders appreciable influence over the supervision of hands and the payment of wages.

The organization of the factory floor in the Slater mills was a reflection of the dominant position of the male householder. Within each department, the supervisory hierarchy came to reflect the hierarchy of the home. All positions of authority, from the second hand to the overseer, were filled by men. Although female labor was predominant in the industrial labor force, no woman filled a managerial position. Like children, women were the subordinates, not the supervisors. The prefactory family hierarchy, in which au-

25. Ibid., Union Mills, vol. 155, Time Book, 1840. Samuel Slater's wage policies recognized and reinforced the traditional position of the male householder as breadwinner and head of his family. While Slater was willing to make certain concessions to householders, he was not eager to meet their demands in the actual amount of money paid to individual workers. The area in which disagreements surfaced most often between Slater and his work force involved wages. Workers were hired initially at whatever price "the demand for labour may compel us to pay" (Samuel Slater and Sons to Nelson Swathland, February 11, 1834, Slater MSS, Steam Cotton Manufacturing Company, vol. 14). When contracts were renewed, Slater tried to limit wage increases or even to reduce wages. "You will be endeavoring to effect agreements with all the families you can, providing it can be done with little or no advance of wages," wrote Samuel Slater to his son John, agent at Webster, on March 27, 1826 (ibid., Samuel Slater and Sons, vol. 235). This statement represented the start, not the end, of negotiations between Slater and individual householders. Both sides stood firm. The weeks of contract negotiations were a sensitive and sometimes a tense period in the relations between Slater and householders. But there was room for compromise in the complex system of wage rates and payment.

thority and power were vested in the husband, was transferred from the home to the new industrial order. The factory system did not challenge paternal authority; it perpetuated it.[26]

In Slater factories, the lesser supervisory positions, such as second hand, were held by young men from eighteen to twenty-two years of age, while overseers were usually men in their thirties and forties. Supervisors normally had no familial ties with the people they directed, but holding the entry-level management position of second hand often entailed the supervision of brothers, sisters, and other relatives. At Union Mills, John McCausland, Jr., aged nineteen, served as a second hand in the carding room, where his younger brother and sister worked; Major Goddard, also nineteen and a second hand in the spinning room, worked with his younger brother and sister. Spinning, weaving, and carding room overseers, however, did not work with kin.[27] This arrangement allowed them to assume a father-surrogate position vis-à-vis all the children and to exercise within the departments the authority nominally vested in parents. Under such conditions discipline proved easy, because children seldom questioned the authority of an overseer or second hand.

The wage scale adopted by Samuel Slater also recognized the customary status held by the householder as primary provider for the family and further enhanced paternal authority. In the 1830s and 1840s rates paid to unskilled and semiskilled male workers ranged from $0.65 to $1.00 per day, more than twice the rates paid to adolescents and unmarried female operatives and three or four times the rates paid to children. Of course, skilled and supervisory staff earned even more. Paid on a piece-rate basis, mule spinners earned from $0.09 to $0.16 per 100 skeins spinning or from $35.00 to $45.00 per month. Dressers were among the highest-paid workers in the factory. At $0.04 per cut, they received approximately $70.00 per month and often much more.[28] Paid by the day, over-

26. William Lazonick, "The Subjection of Labour to Capital: The Rise of the Capitalist System," *Review of Radical Political Economics* 10 (1978):9.

27. Slater MSS, Union Mills, vol. 155, Time Book, 1840; see also Yans McLaughlin, *Family and Community*, pp. 185–195. Among Italian cannery workers discipline remained in the parents' hands, but in Webster direct supervision by parents did not always prevail.

28. Slater MSS, Slater and Kimball, vol. 3, Agreements with Help, 1827–1840; Union Mills, vol. 155, Time Book, 1840.

seers who supervised the carding, spinning, and weaving departments earned somewhat less. Depending on which department they
supervised, their wages fluctuated between $0.83 and $1.50 per
day. (See Table 5.) Wages for other men employed by Slater varied
widely: machinists earned between $15.00 and $20.00 per month,
teamsters received from $23.00 to $27.00 per month, and black-
smiths earned slightly more, $28.00 per month.[29]

Until the 1830s householders received a daily ration of spirits or
wine as part of their payment. This was a cherished custom, one
not easily relinquished by laboring men. For generations the consumption of liquor by workmen and artisans had been an accepted
part of life in New England. During the working day, laborers had
received a ration of gin, rum, or cider brandy, often totaling as
much as a gallon per month.[30] "Please to send by Stephen Buffum
three gallons of brandy for the use of the mill," wrote John Slater
in 1808.[31] In the daybooks kept by the various Slater companies,

Table 5. Daily rates paid to overseers, Slater and Kimball, 1827–1836

Year	Weaving room	Carding room	Spinning room
1827	$1.00		
1828	$1.41		
1829	$1.25–1.33		
1830	$1.03–1.08	$1.08	$0.88
1831	$1.16–1.25		$0.83–0.93
1832	$1.24–1.33	$1.24–1.41	$0.91–1.08
1833	$1.16	$1.08	$0.91–1.00
1834		$1.00	$0.83
1835	$1.50	$1.50	$1.16
1836	$1.33–1.50	$1.50	$1.24

Source: Slater MSS, Slater and Kimball, vol. 3, Agreements with Help, 1827–1836.

29. In all cases, however, these wage rates are only approximations, for Samuel
Slater still calculated his wage bill in British pounds, shillings, and pence. See Slater
MSS, Sutton Manufacturing Company, vol. 11, 1836–1850; Slater and Kimball, vol.
3, Agreements with Help, 1827–1840; Union Mills, vol. 155, Time book, 1840.

30. Adams, *Three Episodes of Massachusetts History*, 2:790; Faler, "Workingmen,
Mechanics, and Social Change," p. 266; Slater MSS, Slater and Tiffany, vols. 3 and
4; vol. 101, January 18, 1827. See also Almy and Brown MSS, John Slater to Almy
and Brown, September 13, 1808; Buck, *Historical Discourse Delivered at the Semi-
Centennial Anniversary of the Slatersville Congregational Church*, p. 10.

31. Almy and Brown MSS, John Slater to Almy and Brown, September 13, 1808.
See also Lease: John Slater and Others to Moses Buffum and Others, Smithfield,
March 1, 1837, S. & J. Slater Collection, Rhode Island Historical Society, Provi-
dence.

the amount of brandy, gin, wine, or rum dispensed to each house-holder was recorded: Cyril Flint, one quart wine; Hanson Bates, one-half gallon rum.[32] Liquor was such an accepted part of the work routine in factory villages that the building of a road, the mending of a fence, or the construction of a building, even a church, would have been unthinkable without the rum ration. In the 1820s, during the construction of the first Methodist church in Webster, the pastor had to purchase several gallons of rum for his workmen. Believing that an explanation was due his congregation, the pastor said "This item of rum was produced for medicinal purpose to be given to the men who were injured when a gale of wind blew off the rafters as the workmen were putting on the roof. Mr. Abbingine Marsh of Charlton had a leg broken and sustained other injuries. Several others were also injured."[33] Under the Slater system, the practice of dispensing liquor to employees eventually perished.

As noted previously, wages paid to child and adolescent factory operatives were markedly lower than those paid to householders. In the carding and the spinning room, rates varied according to the child's age and the length of employment. Over time the minimum wages for these laborers showed an increase. The minimum weekly wages for children in 1796, 1817, 1831, and 1840 were $.0.34, $0.50, $0.60, and $1.00, respectively.[34] Although beginning wage levels were low, they often exceeded those paid by Slater's competitors. In 1838 the Blackstone Manufacturing Company, one of the largest textile concerns in the region, paid children a minimum wage of approximately $0.75 per week, or about $0.25 less than Samuel Slater and Sons paid.[35]

32. Slater MSS, Slater and Tiffany, vol. 4, 1814; vol. 3, 1813. After the practice was discontinued in the 1830s, taverns appeared to meet worker demand for spirits and for a place to meet and relax after the working day. Three taverns competed for the local trade in Slatersville and one in Webster. See Slater and Tiffany, vol. 101, January 18, 1827; Slater and Kimball, vol. 3, Agreements with Help, March 1, 1837; Buck, *Historical Discourse Delivered at the Semi-Centennial Anniversary of the Slatersville Congregational Church*, p. 10.

33. See United Church of Christ MSS, "Methodist Episcopal Church at the Four Corners." n.d.

34. Almy and Brown MSS, account of Roger Alexander, November 1794–April 2, 1796; Slater MSS, Slater and Tiffany, no vol. cited, August 5–December 2, 1820, quoted in Kulik, Park, and Penn, eds., *New England Mill Village*, p. 420; Slater MSS, Slater and Kimball, vol. 3, Agreements with Help, 1827–1840; Union Mill, vol. 155, Time Book, 1840.

35. Blackstone Manufacturing Company Collection, Contract Book, April 1838; see also Slater MSS, Slater and Kimball, vol. 3, Agreements with Help, 1827–1840.

Children who remained with a Slater firm could expect to receive increases in pay, although raises often were small. The wage schedule of Willard Howland's family, shown in Table 6, was representative, and illustrates another feature common to Slater mills. Like Polly and Hannah Howland, many young women who remained employed by Samuel Slater for several years eventually secured jobs as weavers—a semiskilled occupation.[36]

Paid on a piece-rate scale of from $0.20 to $0.25 per cut for four-by-four and seven-by-eight sheeting and shirting in 1840, young female weavers could earn from $16 to $18 per month. The seventeen-year-old Graham twins, Anna and Margaret, consistently received the highest wages paid in the weaving department: $185 and $154 in 1840 for weaving four-by-four cloth. Most weavers, however, received less: from $140 to $150 when they worked a complete year. Work was not always available, however, and many women worked part time: some came in for several hours each day, while others worked for several months at a stretch, stopped for a time, and then resumed work. Lucinda Peck, for example, tended looms from January through July; during this period her monthly wages fluctuated widely, from a low of $6.80 in March to $16.20 the following month. She then left the department, only to return again in November and finish out the year. For 1840 she received a total of $113.80. Sometimes when work was reduced in this department, the women were transferred to another room. Mary

Table 6. Weekly earnings of members of Willard Howland family, 1827–1833

	1827	1828	1829	1830	1831	1832	1833
Malvin	12s.	16s. 6d.	18s.	19s.	21s.	—	—
Munyan	3s. 6d.	4s.	4s. 6d.	5s.	7s.	10s.	12s.
John	3s.	3s. 6d.	4s.	4s. 6d.	6s.	8s.	10s.
Polly	9s. 6d.	12s.	12s.	12s.	—	weave by yard	weave by yard
Hannah	5s.	6s.	8s. 6d.	10s.	12s.	12s.	weave by yard

Note: Later Rebecca, Lorinda, and Paulina joined the work force.
Source: Slater MSS, Slater and Kimball, vol. 3, Agreements with Help, Willard Howland, 1827–1833.

36. Slater MSS, Slater and Kimball, vol. 3, Agreements with Help, 1827–1840; Union Mills, vol. 155, Time Book, 1840.

Archer began the year in the weaving room, where she worked throughout the winter; in May she was transferred to the drawing and trimming section, and she remained there for the rest of the year. Altogether in 1840 she earned $91.60.[37]

METHODS OF WAGE PAYMENT

In no significant area did Slater depart from customary practices and challenge parental authority. His method of wage payment assured householders that parents, not children and adolescents, received wages and had discretion over the disposal of income. Long a tradition in the region, this practice continued into the antebellum period. Under the family wage system, the company credited the wages earned by children and adolescents under age twenty to the factory account of their parents or guardians. Children neither collected nor controlled their own earnings.[38]

Unlike Lowell manufacturers, who paid hands in cash, Samuel Slater paid householders in both cash and kind. One reason was that the banking system remained in considerable disarray for much of the era. Almost all banks in the United States suspended specie payment in 1819 and again in 1837; local circumstances sometimes forced institutions to suspend payment for short periods, as well, and then there was the confusion and inconvenience caused by the plethora of local banknotes in circulation and the discounting that occurred. This was a particular problem for Samuel Slater and his sons. In June 1834 Alexander Hodges, agent of the Wilkinsonville factory, tried to secure funds at the Millbury Bank. He was informed that "they had no funds in any of the Providence or Boston Banks therefore he could not give a check as you requested—nor could he pay it in Providence Bills as he had none &c &c they will except it to pay with their own money and no other."[39] Although in this situation Millbury banknotes might have proved acceptable, some creditors demanded payment in notes from specific regions,

37. Slater MSS, Union Mills, vol. 155, Time Book, 1840.
38. Ibid., vol. 30, Petty Ledger, 1840–1843.
39. Ibid., Samuel Slater and Sons, vol. 236, Alexander Hodges to John Slater, June 27, 1834.

cities, or institutions. In the late autumn of 1839 Francis Fitts of West Woodstock sent a stinging note of complaint to Samuel Slater and Sons on this issue.

> I understand by it that you cannot pay your note in anything but Providence money. This is a fine statement that I must take money that will not pay without a discount of 10 percent even for goods. If you will pay me as good money as you recd of me that is all I ask. I shall be at Providence as soon as the 15 and shall expect the money for your note as I have made arrangements to use it and I must have such as will go at par in New York or Boston.[40]

To solve the problem, Horatio Nelson Slater traveled to New York and tried to trade Providence money. He appraised his brothers of the situation there:

> Providence money is improving in value it is now quick at 3% I have not disposed of more than 600 doll as yet. Mr. Fitts has not yet said that he must now have the money without further delay. It would not surprise me if he should not want it at all—There is no business doing but on the whole a decided improvement in money affairs is apparent. Money is not so excruciating as it was two weeks since. . . . The great cause of fear in our line of business is that the jobbers will have exacted their respectable paper before the December payments which are very heavy.[41]

Slater had reason for concern. Samuel Slater and Sons often had difficulty collecting the debts owed them. Commission agents especially postponed payment, and explained that "we intend to pay it by and by, but at present it is impossible."[42]

During periods of economic dislocation and when it was short of funds, the Slater family solved its financial problems partly at the expense of the workers by postponing wage payments. This often occurred around the annual settlement day of April 1, when workers renewed their contracts. At that time agents required a considerable amount of cash. In 1836 Charles Waite, agent of the Phoenix Thread Mill, for example, requested $6,000 for "paying off the

40. Ibid., vol. 204, Francis Fitts to Samuel Slater and Sons, November 9, 1839.
41. Ibid., vol. 204, H. N. Slater to Samuel Slater and Sons, November 19, 1839.
42. Ibid., vol. 204, O. Wales to Samuel Slater and Sons, March 14, 1838.

help."[43] Sometimes the firm could not meet such demands. In 1839 six or seven weeks elapsed before hands received their annual settlements, and this delay caused considerable friction between factory and firm. Waite complained: "Delay in paying off the help beyond the time it is due engenders a bad state of feeling. They think it is done by their employers to save interest."[44] By the 1840s the situation had eased somewhat, but the firm continued to have liquidity problems throughout the remainder of the antebellum years.[45]

Financial difficulties that recurred throughout the early nineteenth century caused Samuel Slater to seek an alternative, and payment in kind appeared to be a solution. Payment in kind had its roots in a colonial past in which farmhands, day laborers, and artisans received corn, wheat, dry goods, tobacco, rum, land, or farm animals in lieu of money wages. This practice, especially common throughout New England, was transferred to the early factory system.[46]

At Slater's factories, deductions for food, rent, fuel, cattle feed, dry goods, cloth, and various services were made against householders' accounts. Throughout the year workers requested and received cash advances, but the amounts allocated were usually small, between $0.25 and $1.00, although sometimes amounts as large as $10.00 were advanced workers. At the end of March the family account was balanced and, if appropriate, a cash settlement was made or a promissory note issued. Sometimes cash settlements amounted to only a few dollars, and at other times they ran into hundreds of dollars; payment of the larger settlements was spread over a four-to-six-week period. The promissory note with a guaranteed interest rate of 6 percent was preferred by the company.[47]

The ratio of cash and kind payment fluctuated over the years.

43. Ibid., Union Mills, vol. 185, Charles Waite to Samuel Slater and Sons, March 17, 1836.
44. Ibid., Union Mills, vol. 186, Charles Waite to Samuel Slater and Sons, May 17, 1839.
45. Ibid., vol. 189, Fletcher to Union Mills, December 15, 1854. In this note, Fletcher, the Providence agent for the firm, cautioned the factory agent: "Money is very close and it is rather inconvenient to forward this. We hope you will put off the 'evil day' as far as possible and call very lightly when obliged to at all."
46. Richard B. Morris, *Government and Labor in Early America* (New York, 1946), p. 208.
47. Slater MSS, Union Mills, vol. 101, 1837–1849; vol. 110, 1848–1861; vol. 30, Petty Ledger, 1840–1843; vol. 99, Advances, 1851.

In 1820, for example, William Davis and four of his children earned a total of $440.68. Of this amount, approximately 330.00, or 82.5 percent of their income, was spent on food, clothing, rent, fuel, household supplies, and other services or expenses.[48] Later factory account books show that the situation had changed by the 1840s. In 1840–41 families received a significant portion of their income in the form of cash. The income distribution of the John Mc-Causland family (57 percent in kind) was not unusual for the period. (See Table 7.)

Other accounts were more complicated. While children at the age of twenty received their own wages and had their own accounts, many continued to live at home with their parents. In these cases the firm deducted the cost of room, board, and other expenses from their children's earnings and credited the amount to the parent's account.[49] (See Table 8.)

Those who received payment in kind drew against their accounts at the Slater store. In many industrial communities, hands were

Table 7. Income and expenses of John McCausland family, April 1840–March 1841

Income			
	Age	Occupation	Earnings
John McCausland	63	Watchman	$303.40
John McCausland, Jr.	19	Second hand	220.67
Elisa McCausland	10	Carding room	54.47
Extra work			0.99
Total			$579.53

Expenses	
Item	Cost
Rent	$25.00
Cowkeeping	12.74
Cloth	9.20
Potatoes	3.25
Company store	24.50
Other stores	175.87
Cash	328.97
Total	$579.53

Source: Slater MSS, Union Mills, vol. 30, Petty Ledger, 1840–1843.

48. Kulik, Parks, and Penn, eds., *New England Mill Village*, p. 431.
49. Slater MSS, Union Mills, vol. 30, Petty Ledger, 1840–1843.

Table 8. Income and expenses of Benjamin Goddard family attributable to working children, April 1840–March 1841

	Income		
	Age	Occupation	Earnings
Samuel Goddard	14	Spinning room	$63.65
Lydia Goddard	14	Spinning room	57.14
Major Goddard	19	Second hand	169.79
Lucy Goddard	?	Weaver	21.33*
Dolly Goddard	20	Weaver	26.67*
Amy Goddard	20	Spooling	37.33*
Relief Goddard	21	Weaver	12.57*
Total			$388.48

*Amount deducted from earnings and transferred to account of Benjamin Goddard as compensation for room, board, and other expenses of children 20 years and over living with parents.

Expenses	
Item	Cost
Rent	$34.97
Cowkeeping	10.53
Company Store	40.05
Fuel	30.48
L. Berry	110.27
E. White	9.87
Cloth	4.02
Potatoes	7.50
Cash	140.79
Total	$388.48

Source: Slater MSS, Union Mills, vol. 30, Petty Ledger, 1840–1843.

forced to trade with a company store, and prices charged for goods often were considerably higher than elsewhere. At the Crompton Mills in Rhode Island, Benjamin Cozzens openly coerced workers into trading with his establishment. A broadside posted outside his factory read:

NOTICE: Those employed at these mills and works will take notice, that a store is kept for their accommodation, where they can purchase the best of goods at fair prices, and it is expected that all will draw their goods from said store. Those who do not are informed, that there are plenty of others who would be glad to take their places at less wages.[50]

50. John Fitch, *Notes of Travel in the United States*, in *A Documentary History of American Industrial Society*, ed. John R. Commons, 10 vols. (Cleveland, 1910), 7:50.

The Slater company operated a store for its hands in Webster, but it did not force workers to patronize it exclusively. Samuel Slater and Sons leased several grocery and dry goods stores to local proprietors who agreed to sell goods to employees on account and to accept quarterly settlements from the company. Hands were also permitted to trade with any shopkeeper who accepted their accounts. In 1832 twelve shops operated in Webster: five grocery and dry goods stores, three general stores, a hardware store, a bakery, a boot and shoe shop, and an establishment that sold drugs and liquor. Still, the company store absorbed a significant proportion of their wages. Workers received due bills drawn on either one of the Slater company stores or one of the local shopkeepers. Issued every four weeks, these slips listed the householder's name, the amount due to him for his labor and for that of his children and charges, and the date. On presentation of the chit to the appropriate store clerk, the householder was given whatever groceries and dry goods he desired. But the entire amount of the chit was seldom spent at one visit. Employees drew goods every few days. The clerk retained the chit and listed all the goods taken, and at the end of the four-week period the chit was forwarded to the mill office and a settlement was made.[51]

Economic survival in Webster and Slatersville, however, was not completely dependent on factory wages. Those who rented factory cottages received an allotment of land large enough to grow a variety of spring and summer vegetables. For a fee of 26 cents the firm plowed the garden for families. Much of the meat, poultry, eggs, milk, and other dairy products consumed by factory families came from the farm animals they brought with them when they migrated to the industrial towns. While a charge of 25 cents per month per animal was assessed against those householders who grazed their horses, cattle, and dairy cows on company pasture,

51. "Two Hundred Years of Progress," p. 10. See also Slater MSS, Samuel Slater and Sons, vol. 236, E. Knight to J. Slater, February 12, 1836; Union Mills, vol. 30, Petty Ledger, 1840–1843; vol. 195, M. Lyon, May 25, 1837; vol. 101, 1837–1849, Samuel Slater and Sons and John Dixon and Sons, agreement, August 19, 1837; vol. 195, South Village Store, September 21, 1837; October 9, 1837; November 1, 1837; vol. 179, Webster Woolen mills to Samuel Slater and Sons, September 13, 1850; vol. 195, Merchandise Account at South Village Store, October 1, 1838; vol. 87, Rent Book, Dixon, April–November 1849; Otis Foster, April 9, 1849; Moses Phipps, February 11, 1851; and Ware, *Early New England Cotton Manufacture*, p. 245.

few restrictions were placed on the number of smaller animals and birds they could keep in or near company dwellings.[52]

MANAGEMENT'S PREROGATIVES

The conditions of labor observed in Slater's communities represented a compromise between the demands of householders and the requirements of the new production system. Within the new industrial order, the traditional status of the male and female householder as provider and protector of the family was preserved. By providing householders with socially acceptable forms of task-oriented labor, by allowing parents appreciable influence over the conditions under which their children worked, and by paying householders all wages earned by their families, Slater eased the transition of hundreds of families from farm to factory. Yet their control was limited.

Negotiations between householder and management encompassed a variety of issues, but one area in which factory families had little influence was the scheduling of work. Samuel Slater set the operating schedule of the factory. In the 1820s and 1830s hands worked six days each week, from twelve to fourteen hours each day. During the winter months operations began at daylight and continued until 8:00 P.M., while during the summer laborers worked from 5:00 A.M. until 7:00 P.M. Within the mill the long workday was broken by two meal breaks, one lasting approximately thirty minutes and another between thirty and forty-five minutes. The factory bell, the only time mechanism found in the mills, signaled the beginning and the end of each meal break, summoned hands to the factory in the morning, and tolled the end of the workday.[53]

One of the chief drawbacks to the new factory system and an-

52. Slater MSS, Union Mills, vol. 101; Sutton Manufacturing Company, vol. 5, Journal A, June 26, 1830.
53. Slater MSS, Sutton Manufacturing Company, vol. 45, October 1832; Union Mills, vol. 155, Time Book, 1840; Slater and Kimball, vol. 3, Agreements with Help, 1827–1839. See also *Paterson Courier*, August 11, 1835, quoted in Commons, ed., *Documentary History of American Industrial Society*, 5:64. For the Lowell situation see Josephson, *Golden Threads*, pp. 219–220; and Harriet Hanson Robinson, "Life of the Early Mill-Girls," *Journal of Social Science* 16 (December 1882): 129.

other area in which householders exercised little influence was unemployment. The time and wage records for the Phoenix Thread
Mill and the Union Mills from 1839 to 1844 reveal that the comany
closed down the factories completely on some days and worked a
part-time schedule on others. No seasonal pattern is evident. At
the Phoenix Thread Mill in 1839, for example, all hands worked
part time on August 31 and on December 16, 17, 18, 21, and 23.
During the following year the Union Mills operatives worked thirty-
eight short days. While the severe economic downturn of the period might explain the labor schedules for 1839 and 1840, the situation did not improve when the industry began to revive a few
years later. In 1844 the Phoenix Thread Mill closed down completely for several weeks, first from April 1 through April 6 and
again from August 26 through August 31. On May 7, hands worked
a nine-hour day. This intermittent, irregular work pattern persisted during the antebellum period.[54]

Even when the factory was open and operating, jobs were not
available to all who wanted work. During the depression year of
1840, periodic unemployment was a feature of factory life. At Slater's Union Mills, for example, unskilled hands worked on average
136 days. Albert, Sabrina, Roxana, and Abegail Drake worked 167,
154, 106 and 86 days, respectively. Skilled and semiskilled hands
fared better: they worked approximately 182 days that year. Even
supervisory staff put in less time than usual. John Leavins, overseer
in the cloth room, worked 282 days, while Rufus Freeman, overseer of the dressing, spooling, trimming, and warping section,
worked only 200 days.[55] As Calvin Colton noted, that year was a
particularly difficult one for labor: "Everybody knows that, in 1840,
labor went begging for bread, and could not always get it."[56] Although prosperity returned to the industry several years later, hands
still worked an irregular schedule. For the six months ending June
30, 1844, the Phoenix Thread Mill operated for 137 days, but fully
52 percent of the operatives with jobs in the spinning and the card-

---

54. Slater MSS, Union Mills, vols. 144, 145, 149, 155, 1836–1866; see also Phoenix Thread Mill, vols. 26, 27, 1839–1850 (all time books).
55. Ibid., Union Mills, vol. 155, Time Book, 1840.
56. Calvin Colton, *Public Economy for the United States* (New York, 1969 [1848]), p.
411.

ing rooms worked fewer than 70 days during that period. While some of these hands undoubtedly were transferred to other rooms or sections of the factory, many were laid off.[57]

Personal reasons such as illness, inclement weather, and attendance at religious or political functions cut into work schedules and further reduced the amount of time laborers spent in the factories. Among those who worked with some degree of regularity throughout the year, most were absent for at least one or two days each month.[58]

Intermittant employment, however, was not new. In other economic pursuits, factory families also faced economic uncertainty. Agricultural work was seasonal and domestic industry proved unreliable. Demand for palm-leaf hats, gloves, shoes, and other fashion items manufactured in the home fluctuated widely, and few people worked steadily. Families survived by combining income from all sources, by relying on the food they produced, and by turning to the church and to family or friends for assistance.

Despite the drawbacks in factory employment, Slater attracted family labor to his mills. In the organization and operation of his industrial communities, Slater respected the desires of householders and incorporated traditional values, practices, and customs within the new order. In return for his safeguarding of traditional prerogatives, householders provided Slater with a steady supply of tractable, industrious, reliable hands. The bargain that had been made between Samuel Slater and his workers in Pawtucket could be seen operating in Slatersville, Webster, Wilkinsonville, and the many other Slater-style communities throughout New England. As long as both labor and management adhered to its side of the agreement, harmonious relations between the two parties prevailed.

57. Slater MSS, Phoenix Thread Mill, vol. 28, Time Book, 1839–1850.
58. N. B. Gordon, Agent's Diary, March 24 and July 13, 1839; March 1, 1830; January 29, 1839; Mansfield U. C. & W. Manufacturing Company Manuscripts, Baker Library, Harvard University. See also Slater MSS, Phoenix Thread Mill, vol. 28, especially records for Mary Raymond and Joel Foster; H. N. Slater Papers, vol. 33, Alexander Hodges to H. Slater, September 12, 1839.

[7]

Factory Discipline

Industrial discipline posed some of the gravest problems faced by early factory masters, who had to devise various methods to teach people the so-called habits of industry: regularity, obedience, sobriety, steady intensity, and punctuality. Most manufacturers solved such problems in one of two ways: the stick-and-carrot approach as described by Sidney Pollard or the more subtle, internalized form of discipline as discussed by E. P. Thompson. Using the arguments of Max Weber and Erich Fromm, Thompson has maintained that rewards and punishments do not always succeed in creating a disciplined, tractable, steady, industrious laborer. Internal forces or "inner compulsion," he asserts, usually prove more "effective in harnessing all energies to work than any other compulsion can ever be."[1] For this method to be effective, each worker had to be made his or her own taskmaster, had to be made to feel guilty for "deviant" conduct, and had to develop an internal drive toward right and proper behavior. Among the British working class, Thompson ascribed that internal force to religion, especially Methodism. In Webster, Slatersville, and Wilkinsonville internal self-control also served to discipline the labor force. There the church and the family were the twin forces employed to exact compliance with factory rules and regulations. Discipline, however, did not begin and end at the factory door; beliefs taught in the church and the home circumscribed the behavior of the entire society.

1. Edward P. Thompson, *The Making of the English Working Class* (London, 1964), pp. 357–358.

In the United States and Britain the rise of evangelical churches coincided with the growth of the factory system. In both countries the church and its many activities received considerable support from local manufcturers. In Slatersville, the Congregational church dominated the religious life of the community. John Slater built the church, paid for its continued maintenance, became involved in all church activities, and encouraged his work force to join the congregation. Writing in 1866, the Reverend E. A. Buck testified that John Slater "gathered around him a highly worthy class of laborers, many of them of a decidedly religious character."[2]

In Webster the Baptists, the Methodists, and the Congregationalists all received support from the Slater family, and although Samuel Slater and his sons belonged to the Congregational church, the Methodists in Webster received favored treatment. Samuel Slater encouraged the establishment of the Methodist church, and throughout the antebellum period his company continued to support it. Slater provided a plot of land for the church, laid its foundation, and bought sixteen of its forty pews for the exclusive use of his employees. Slater's company later built the parsonage and was responsible for the continued maintenance of church buildings. Whenever church officials required financial assistance, they turned first to the company and only later to the congregation. In effect the company became a church proprietor, exercising considerable influence over religious matters.[3]

The Slater family encouraged its laborers to attend worship services and participate in other church activities. John Slater, Sr., expected all company employees to attend Sabbath services: "It has long been one of the established regulations of the mills, that the help are expected to attend public worship on the Sabbath. Also that no work will be done or repairs made by the company on that day."[4] Manufacturers also allowed laborers leave to attend special

2. Buck, *Historical Discourse Delivered at the Semi-Centennial Anniversary of the Slatersville Congregational Church*, p. 7; Donnell, *Historical Address Delivered at the Centennial Celebration of the Congregational Sunday School*, pp. 16–17, 19–20.

3. Slater MSS, Samuel Slater and Sons, vol. 235, Samuel Slater to John Slater, January 24 and 27, 1834; Phoenix Thread Mill, Asher Joslin to Storrs, June 21, 1847; see also United Church of Christ MSS, Records of Board Meetings for Webster Episcopal Church, September 12, 1853, August 7, 1854; and "Methodist Episcopal Church at the Four Corners," September 18, 1832.

4. Buck, *Historical Discourse Delivered at the Semi-Centennial Anniversary of the Slatersville Congregational Church*, p. 24.

church functions, often shutting down the mills and closing their shops so that all hands could attend revivals, camp meetings, and special quarterly sessions. During the summer revival of 1839, the mills were closed for several days to allow local Methodists to attend the meeting. "This being camp meeting with our Good Methodists," wrote factory agent Alexander Hodges, it "will be rather a broken one with the Mills."[5] Clearly, Methodists formed such a large proportion of the labor force that their absence effectively curtailed operations.

By the late 1830s and early 1840s, Slater's labor force was predominantly Methodist. A comparison of employment ledgers and local church membership rolls reveals that prominent members of church boards, superintendents of Sunday schools, lay preachers, and stewards held important skilled and supervisory positions in local Slater factories. Charles Waite, resident manager of Slater's Phoenix Thread Mill, served variously as Sabbath school administrator, treasurer, and secretary-treasurer of the Methodist church. William Kimball, for ten years resident superintendent of the Slater and Kimball mill, served as the assistant secretary of the Methodist Sabbath school. At Union Mills, four of the six machinists, two of the four dressers, all of the bailers, and many of the mule spinners belonged to the church. Their sons and daughters filled the unskilled jobs in the carding, spinning, and weaving rooms.[6]

This pattern developed throughout the region as manufacturers invested heavily in religious organizations. They donated land, built churches, paid the pastors' salaries, subscribed to choral societies, and sometimes purchased pews. In neighboring Sutton Township, for example, textile proprietors in Wilkinsonville and Manchaug villages not only helped to construct the local churches, but also paid the minister's salary. In South Sutton, Welcome Whipple, proprietor of the Douglas Cotton Manufacturing Company, built two churches, one Methodist and the other Baptist, and partially financed their continued maintenance. Manufacturers in Blackstone, a community situated near the border between Rhode Island and Massachusetts, went even further. In addition to building

5. Slater MSS, H. N. Slater, vol. 33, A. Hodges to H. Slater, September 12, 1839.
6. Ibid., Union Mills, vol. 155, and Slater and Kimball, vol. 3, Agreements with Help, 1827–1839; United Church of Christ MSS, "Constitution of the Methodist Episcopal Church Sabbath School," 1830–1840.

and maintaining the local church, they deducted pew fees directly from their employees' salaries in much the same manner as union dues are deducted today.[7] Support for religion was not confined to Slater-style manufacturers. In Waltham by 1837 three Congregational churches, one Methodist church, and a Universalist church had been established. In Lowell the manufacturers constructed or subsidized eight churches and mandated that "a regular attendance on public worship on the Sabbath is necessary for the preservation of good order."[8] Some manufacturers went further and tried to make religion a criteria for employment. Jedediah Tracy, a manufacturer in Troy, New York, advised on December 27, 1827: "It should be the first object of our manufacturing establishments, to have their superintendents, and overseers, and agents, men of religious principles, and let it be felt by the owners that it is always for their interest to support religion."[9]

THE SUNDAY SCHOOL, HOME DISCIPLINE, AND THE WORK ETHIC

In part Samuel Slater and other manufacturers supported religion because they viewed it as a form of social control which facilitated the discipline of workers. The dictums and discipline advanced by the church became part of the foundation of a work

7. William A. Benedict and Hiram A. Tracy, *History of the Town of Sutton, Massachusetts, from 1704 to 1876, including Grafton until 1735; Millbury until 1813; and Parts of Northbridge, Upton, & Auburn* (Worcester, 1878), p, 485. Emerson, *History of the Town of Douglas*, p. 240; Buck, *Historical Discourse Delivered at the Semi-Centennial Anniversary of the Slatersville Congregational Church*, pp. 8, 24; Hurd, *History of Worcester County*, 1:611; Steere, *History of the Town of Smithfield*, pp. 104–105. Labor Books, 1853–1854, Blackstone Manufacturing Company Collection. John Slater contributed to other churches throughout the region. In 1833 and again in 1834 he donated more than $100 to St. James Methodist Church, Woonsocket Falls. See Slater MSS, S. & J. Slater, Philip B. Stiness to John Slater, May 18, 1833, and Deed for Pew, September 29, 1834. George Slater contributed to the local Baptist church in Webster. See Sutton Manufacturing Company, vol. 45, George Slater in Account with Union Mills, April 8, 1837; May 18 and June 24, 1838.

8. "Regulations to be Observed by all Persons Employed by the Suffolk Manufacturing Company," n.d., quoted in Dublin, *Women at Work*, p. 78; Daniel W. Howe, *The Political Culture of the American Whigs* (Chicago, 1979), p. 1.

9. Letter of Jedediah Tracy, December 27, 1827 (cited in White, *Memoir of Samuel Slater*, p. 132).

ethic, and as such they served to train, discipline, and control workers.[10] This was the case in Webster, where the Methodist church educated a whole generation in the dictates of their religion. The written tracts, hymns, sermons, and other literature used in the church all advanced the same messages: obedience, deference, industry, honesty, punctuality, and temperance. The lessons prepared the young operatives for ultimate salvation and also trained them to be good, obedient factory hands. In Webster the Methodist Sabbath school was the principal agency through which these values were taught.[11]

The Webster Sabbath school flourished. In 1841 A. D. Merrill, minister of the local church, reported:

> The present state of the Sabbath School in the Webster Station of the M. E. Church must be viewed as in a state of more than ordinary prosperity. . . . It is the sentiment of the Superintendant that the school is more prosperous now than ever before at this season of the year since he resided in town. Such is the interest felt by the Superintendants and Teachers that they have during the last Quarter established a monthly Prayer meeting for the benefit and spiritual interest of the school.[12]

This Sabbath school owed its origin to Samuel Slater. Like his mentor, Jedediah Strutt, Samuel Slater established Sabbath schools in each of his industrial villages. Based on the British system he had observed, these schools were to "condition the children for their primary duty in life as hewers of wood and drawers of water."[13]

10. In Britain Methodism served to control and discipline workers. See Elie Halevy, *A History of the English People in the Nineteenth Century* (London, 1924), vol. 1, *England in 1815*; see also E. J. Hobsbawm, *Labouring Men: Studies in the History of Labour* (Garden City, N.Y., 1967); Thompson, *Making of the English Working Class*, pp. 350–400; Laqueur, *Religion and Respectability*, pp. 78–83.
11. Gilbane, "Social History of Samuel Slater's Pawtucket," pp. 301–313; Donnell, *Historical Address Delivered at the Centennial Celebration of the Congregational Sunday School*, pp. 19–20.
12. United Church of Christ MSS, Minutes of the Webster Quarterly Conference, Report, December 11, 1841.
13. White, *Memoir of Samuel Slater*, p. 107; Fitton and Wadsworth, *Strutts and the Arkwrights*, pp. 102–103; Pinchbeck and Hewitt, *Children in English Society*, 1:293–296; Pollard, *Genesis of Modern Management*, pp. 180–181; Jones, *Charity School Movement*, pp. 142–146; Wadsworth, "First Manchester Sunday Schools," pp. 299, 300, 305; Laqueur, *Religion and Respectability*, pp. 187–238. Laqueur challenges E. P. Thompson's interpretation of the role of the Sunday school in the life of England's working class.

Through these Sunday schools, Samuel Slater sought to foster attitudes toward right and proper conduct that would make children good citizens and good workers. A hymn from Dr. Isaac Watts's songbooks sung by the children in Slatersville and Webster began:

> Why should I deprive my neighbor
> Of his goods against his will?
> Hands were made for honest labour,
> Not to plunder or to steal.[14]

When churches became firmly established in Webster and Slatersville, Slater disbanded his Sunday school and relinquished moral education and industrial training to the churches.

In transferring moral education to religious bodies, Slater could be confident that the church would continue to inculcate virtues and beliefs sympathetic to the new industrial order. In Webster, for example, the men who ran the Sunday school were the same men who supervised local factory operations. For twelve to fourteen hours each day, six days each week, children worked under Charles Waite and William Kimball, and on the seventh day they listened while the same men interpreted the scriptures.[15]

Webster was not unique here. In Slatersville, Amos D. Lockwood, resident agent of the local Slater mill, served as the Sabbath school superintendent and as leader of a Bible class. One of his pupils described his superintendency in the following terms: "The Slatersville Sunday school was an orderly one. How could it be otherwise with Mr. Taylor and Mr. Lockwood, at the helm? When the service was over Mr. Lockwood, who was superintendent, used to stand in the front gallery and call the name of the teacher of each class to pass out, and there was no crowding or noise."[16]

Values taught in the Sunday schools proved favorable to factory

14. See Isaac Watts, *Divine and Moral Songs for Children* (London, n.d.), p. 89. For a list of the materials found in one of Slater's schools, see Bremner, ed., *Children and Youth in America*, 1:178; Donnell, *Historical Address Delivered at the Centennial Celebration of the Congregational Sunday School*, p. 37; White, *Memoir of Samuel Slater*, p. 117.

15. Slater MSS, Sutton Manufacturing Company, vol. 45, October 1832; Union Mills, vol. 185, Charles Waite to Samuel Slater and Sons, October 24, 1836; Slater and Kimball, vol. 3, Agreements with Help, 1827–1839; also United Church of Christ MSS, "Constitution of the Methodist Episcopal Church, 1830–1840."

16. Donnell, *Historical Address Delivered at the Centennial Celebration of the Congregational Sunday School*, pp. 31 and 25, 29–30.

discipline. One of the first lessons taught to children concerned obedience. This was the first law of childhood, the first rule of the church, and the regulation deemed indispensable for the smooth, efficient operation of the factory. Sabbath school teachers stressed this dictum and condemned all children who disobeyed those in authority, whether at home, at school, or in the factory. One lesson used to instill this particular value might have been introduced to children in the following way: "As you sit here now, listening to me," the Sabbath school teacher might begin,

> can you remember any disobedient habits of yours, that make the father and the Mother unhappy, when they look at you and see how fast you are growing, without growing better? Is it true that you have had a bad temper, and do not love to be controlled? . . . Is it true that you have grown, but have not grown out of any of these habits: just as bad as ever, just as disobedient, just as wicked with the tongue as ever?[17]

Punctuality, a cornerstone of any work ethic, also received considerable attention from the Methodists. They were concerned about time. In an era when people were accustomed to family time and to task-oriented labor, Methodist children were being taught: "Be punctual. Do everything exactly at the time."[18] In his reports to the company, factory agent and Sabbath school administrator Charles Waite often stressed the need for punctuality: "Punctuality is the life of business whether in the counting house or the factory."[19] The severe style of life demanded of the faithful allowed no place for carefree play, laughter, or harmless pranks. "No room for mirth or trifling here," began a child's hymn on amusement titled "And Am I Only Born to Die?"[20] Children were constantly warned that "life so soon is gone," that although

> We are but young—yet we must die,
> Perhaps our latter end is nigh.[21]

17. *Sunday School Gazette*, Worcester, Mass., October 27, 1849, p. 13.
18. Abel Stevens, *History of the Methodist Episcopal Church in the United States*, 2 vols. (New York, 1864), 2:226–228, 2:231.
19. Slater MSS, Union Mills, vol. 186, Charles Waite to Samuel Slater and Sons, May 17, 1839.
20. *Hymns for Sunday-Schools, Youth, and Children* (New York, 1850), p. 84.
21. *Hymns for Sunday Schools* (New York, 1842), p. 65.

All hymns carried a similar warning: children should "sport no more with idle toys, and seek far purer, richer joys," devote themselves totally to Christ, and obey the teachings of the church.[22]

Many values, including punctuality, attention to duty, and seriousness of purpose, were neatly summarized in the Webster Sabbath school constitution, which was drawn up by local church officials. The constitution was in fact a code of conduct similar to that maintained in the factory. In part the constitution required all children "to be regular in attendance, and punctually present at the hour appointed to open school. To pay a strict and respectful attention to whatever the teacher or Superintendent shall say or request. To avoid whispering, laughing and any other improper conduct."[23] Altogether these values became the moral foundation for a strict work ethic. But one element essential to the successful operation of this ethic was missing: internal self-discipline.

As a work ethic these dictums and values would have been much less effective had not the church also taught self-discipline, self-restraint, and self-regulation of behavior. All efforts were made to internalize values in order to create an inner discipline that would control and limit the child's behavior. Children were taught to "do good" instinctively and to develop an internal drive toward right and proper conduct. In effect, they became their own taskmasters; conscience rather than rewards and punishments directed their actions.

To achieve this end, Sunday school teachers linked proper conduct to grace or, to put it conversely, disobedience to damnation. An exchange used to close an infants' class made the connection explicit:

Teacher: Do you know who belong to Satan's army? Say after me—
All who tell lies, all who swear and cheat; all who steal; all who are cruel.[24]

22. *Sunday School Gazette*, Worcester, Mass., October 27, 1849, p. 16.
23. United Church of Christ MSS, "Constitution of the Methodist Episcopal Church Sabbath School," 1861–1863.
24. Adie Wardle, *History of the Sunday School Movement in the Methodist Episcopal Church* (New York, 1918), p. 110. A verse used at the Slatersville Sabbath school was: "It is a sin to steal a pin, much more a greater thing!" (Donnell, *Historical Address Delivered at the Centennial Celebration of the Congregational Sunday School*, p. 37).

In the child's mind, to cheat, to steal, to lie, or to misbehave in any
way was to violate God's law, lose grace, and risk damnation. And
such a risk was unthinkable. The songs published in *Hymns for Sunday Schools* describe hell in forceful and emotional terms:

> There is a dreadful hell,
> And everlasting pains;
> There sinners must with devils dwell,
> In darkness, fire, and chains.[25]

Images of everlasting punishment and fears of eternal damnation
worked to ensure a strict and steady compliance with the values
advanced by the church.

The Sabbath school trained Webster's child workers well. In the
factory, children quickly learned to obey all orders, for to disobey
was to feel anxious and to risk censure or eternal damnation. Corporal punishment, fines, and the ultimate discipline—dismissal—
were largely absent, and in fact were unnecessary when children
readily and willingly, not to say cheerfully, obeyed the dictates of
second hand and overseer. Operations almost always ran smoothly.
Supervisors faced few disciplinary problems such as absence, theft,
inattention to duties, or general mischievous behavior.[26] The instructions children received in the Sabbath school were largely responsible for the exemplary behavior.

The tenets of the church were reinforced by lessons learned in
the home. In the area of discipline the responsibility of parents was
widely recognized.[27] As Walton Felch, agent of a Medway, Massachusetts, factory, cautioned parents as early as 1816:

> O, anxious parents! train your rising youth,
> In all the FAITHFUL ELEGANCE of Truth;
> Lest, where paternal care has failed to gain,
> A dread futurity the wretch restrain.

25. *Hymns for Sunday Schools*, pp. 91–92.
26. Slater MSS, Slater and Kimball, Agreements with Help, 1827–1840, vol. 3.
27. Louisa Hoare, *Hints for the Improvement of Early Education and Nursery Discipline*
(Salem, 1826), p. 27; Theodore Dwight, Jr., *The Father's Book of Suggestions for the
Government and Instruction of Young Children on Principles Appropriate to a Christian
Country* (Springfield, 1835), p. 207; Humphrey, *Domestic Education*, p. 126; L. Sigourney, *Letters to Mothers* (New York, 1845), p. 58.

Obedience teach; the base whereon thy skill
May raise "high towers" and mighty schemes fulfil;
But mark the means that to the end conduce,
And frame them fit, at least—for mortal use.

· · ·

Not pride—but Prudence, to your offspring teach,
To grasp but honors justly in their reach;
The heights of life with careful steps to scale;
Or walk submissive through its lowly vale;
The suffering good with soothing hand to greet;
To shun the bad, and yet humanely treat;
To love and pity those who are unkind,
But mourn the folly that misleads the mind,
For such may all the living kindly greet,
Tho' free from wicked projects or deceit.[28]

In communities such as Webster and Slatersville, the home became another training ground for a generation of factory hands. Lessons taught there stressed the implicit, unquestioning obedience and deference to authority deemed necessary for good family government, for a well-ordered society, and for the successful operation of the factory system. In antebellum New England familiar values, expressed through child-rearing practices and reinforced by religious tenets, operated to create tractable hands.

Throughout New England, Calvinism was still influential in the sphere of child training. Although traditional precepts and doctrines had undergone substantial transformation by the 1830s, something of the Calvinist emphasis on internal self-control survived. As E. P. Thompson would have it, "inner compulsion" toward right and proper conduct exemplified the New Englander's approach to child training. In nineteenth-century America a "good child" was still defined as one who "shunned the morally wrong because he undeviatingly sought the morally right, and . . . internal sources produced this action so that goodness reigned in both character and conduct."[29] H. Humphrey counseled parents to "enlist the consciences of their children, to secure a ready and cheerful

28. Walton Felch, *The Manufacturer's Pocket-Piece; or, The Cotton-Mill Moralized, a Poem, with Illustrative Notes* (Medway, Mass., 1816), pp. 21–22.

29. Peter G. Slater, "Views of Children and of Child Rearing During the Early National Period: A Study of the New England Intellect," Ph.D. dissertation: University of California, 1970, 219.

obedience. Indeed, till you reach the conscience, you have done but little to bind your child to his duty."[30]

The values taught in the home were those required by industry: they served to make both dutiful, respectful children and submissive workers. Even the most liberal authorities on child-rearing practices, such as Lydia M. Child, cautioned parents that "implicit obedience is the first law of childhood," that "whatever a mother says always must be done."[31] Other writers concurred. John Abbott, described by one historian as the Spock and Seuss to the people of the Civil War generation, went a step further and joined disobedience with wickedness. In *Child at Home* he wrote: "Think you, God can look upon the disobedience of a child as a trifling sin? . . . It is inexcusable ingratitude."

> The only path of safety and happiness is implicit obedience. If you, in the slightest particular, yield to temptation, and do that which you know to be wrong, you will not know when or where to stop. To hide one crime, you will be guilty of another; and thus you will draw upon yourself the frown of your maker, and expose yourself to sorrow for time and eternity.[32]

All commands had to be immediately and cheerfully obeyed. Children were expected to respond to orders with glad and happy hearts. Again John Abbott: "Obedience requires of you, not only to do as you are bidden, but to do it with cheerfulness and alarcity"; and Theodore Dwight, Jr.: "Children should be obedient—must be obedient, habitually and cheerfully."[33]

If obedience was the first law of childhood, then deference was the second. Children quickly realized their subordinate position within the patriarchal family. Mother taught that father was the head and ruler of the household, that he stood before them as

---

30. Humphrey, *Domestic Education*, p. 47. See also Hoare, *Hints for Improvement*, p. 45; John Abbott, *Child at Home; or, The Principles of Filial Duty* (New York, 1833), p. 86.

31. Lydia M. Child, *The Mother's Book* (London, 1832), pp. 25, 50.

32. Abbott, *Child at Home*, pp. 63, 68–69. Two chapters of this seven-chapter book were devoted to the subject of obedience. See also Daniel Calhoun, *The Intelligence of a People* (Princeton, 1973), pp. 156–165.

33. Abbott, *Child at Home*, p. 86; Dwight, *Father's Book of Suggestions*, p. 127; see also Sigourney, *Letters to Mothers*, p. 33; Calhoun, *Intelligence of a People*, pp. 156–165.

God's representative on earth, and that, as supreme earthly legislator, he exercised complete control over their every action. According to Humphrey, "children must early be brought under absolute parental authority, and must submit to all the rules and regulations of the family during the whole period of their minority, and even longer, if they choose to remain at home."[34] Once again religious injunctions were employed. Humphrey warned: "Now to disobey your parents, is to dishonor them. This you have done, and in doing it, you see you have broken God's holy law. We can forgive you, but that will not lessen your guilt, nor procure forgiveness from your heavenly Father. You must repent and do so no more."[35]

Lessons taught in the home, reinforced by tenets learned from the scriptures, became the moral foundation for a disciplined labor force. Workers found little difference between disciplinary patterns in home and factory. Both home and factory were paternalistic, and both were controlled by men who expected unquestioning compliance with all commands. Children merely transferred their values and behavior patterns from the home and the church to the factory; old values were easily accommodated by the new institution.

COMMUNITY DISCIPLINE

Familial and religious doctrines and discipline served as the basis for a well-ordered society. During the first half of the nineteenth century in Webster, overt forms of authority and control were limited. One or two unpaid constables were appointed. Their responsibilities were confined largely to such mundane tasks as restraining "horses, meat cattle and swine from running at large within the limits of the town."[36] All three of the constables who served in Web-

34. Humphrey, *Domestic Education*, p. 41; Dwight, *Father's Book of Suggestions*, p. 119.

35. Humphrey, *Domestic Education*, p. 150; Slater, "Views of Children and of Child Rearing," p. 161; and L. Sigourney, *Letters to Young Ladies* (New York, 1845), p. 246.

36. See Webster, Mass., Town Meetings, Minutes: April 2, 1832, and April 7, 1834; see also Webster, Mass., Town Hall, "List of Constables," 1832–1860; Manuscript Schedules, Sixth Census of the United States, 1840; Slater MSS, Slater and Kimball, vol. 3, Agreements with Help, William Andrews, 1830–1833; United Church of Christ MSS, Constitution of the Methodist Episcopal Church Sabbath School, April 19, 1830–1833.

ster in the 1830s were members of the Methodist Church. Methodists were dominant in the office of town selectmen as well as in the police force. (See Table 9.)

Control of local citizens, however, rested more with religious and family government than with legal authorities. People were expected to regulate their own behavior and to report misconduct to the proper authorities, usually ecclesiastical societies. The churches in Webster provided guidelines for proper behavior, kept up a constant surveillance of society, and brought suspected backsliders before the congregation for trial and punishment. Parishioners not only discovered and prosecuted crimes but also worked to prevent their commission.[37]

In the area of self-discipline, Methodists took the lead. According to the 1832 *Doctrines and Discipline of the Methodist Episcopal Church,* those in good standing had to avoid:

Drunkenness: or drinking spiritous liquors, unless in case of necessity.
Fighting, quarrelling, brawling, brother going to law with brother.
The buying or selling goods that have not paid the duty.
Uncharitable or unprofitable conversation: particularly speaking evil of magistrates or of ministers.
The putting on of gold and costly apparel.
The singing those songs, or reading those books which do not tend to the knowledge or love of God.

Table 9. Religious affiliation of town selectmen, Webster, 1832–1860

Term	Methodist	Congregationalist	Baptist	Unknown
1832–1839	10	2	0	6
1840–1849	7	7	1	2
1850–1859	9	2	0	4

Note: Most of the Congregationalists elected in the 1840s served only one or two years.
Source: Webster, Mass., Webster Town Hall, "List of Selectmen," 1832–1860.

37. See Webster, Mass., Webster Town Hall, "List of Selectmen," 1832–1860. See also United Church of Christ MSS, Minutes of the Webster Quarterly Conference, 1843–1856; "Methodist Episcopal Church at the Four Corners"; Congregational Church Parish Records, 1838–1911.

Softness and needless self-indulgence.

Laying up treasure upon earth.

Borrowing without a probability of paying; or taking up goods without a probability of paying for them.[38]

Voluntary compliance with these rules was expected, but for the vulnerable and the backslider, vigorous measures were taken to enforce adherence. Church members closely monitored the actions of their neighbors and reported to ecclesiastical authorities those who deviated from acceptable norms. Accusations against friends and neighbors ranged from adultery, intemperance, indebtedness, and theft to profanity, malicious gossip, consorting with dubious individuals, and brawling. Accusations were taken seriously, and many cases were brought before the church board or the congregation for settlement.[39]

Some issues absorbed the attention of the church for weeks and even months. Allegations of adultery brought by several fellow Methodists against the widow Harriet Emerson and William Johnson were investigated thoroughly by a specially appointed church committee, which found that

> after calling on different individuals who it was "said" were knowing to the facts stated in the reports and ascertaining that they knew nothing of the matter only as they heard the reports from others, and also ascertaining that said Harriet was at her home on the night it is alledged she spent with said Johnson at his place of labor, and finally, after conversing with the persons whom it is believed started the report, and they affirming that they had never seen any thing criminal in their conduct, or knew any thing injurious to their moral or religious character,—Therefore your committee are of the opinion, that said reports are without any foundation in truth either in whole or in part, and that the accused are still deserving the Christian sympathy of the church and confidence of the Public.[40]

38. *The Doctrines and Discipline of the Methodist Episcopal Church* (New York, 1832), p. 78.

39. United Church of Christ MSS, Records of Board Meeting for Webster Methodist Episcopal Church, 1834–1867; see also Congregational Church Parish Records, 1838–1911.

40. United Church of Christ MSS, Records of Board Meetings for Webster Methodist Episcopal Church, February 1, 1855. The Methodist church was not the only

Most of the disciplinary actions instituted by the Methodists were for infractions much less serious than adultery. The church arbitrated disputes between husbands and wives and between neighbors. When a Methodist refused to pay his or her debts promptly, the church also became involved. A dispute between the Sayles family and John Rickets concerned a debt. Francis Sayles, aged twenty-six and a local merchant, owed money to John Rickets. Rickets hired the mother of Francis Sayles to perform some work for him, but rather than pay her for her labor, Rickets deducted the amount owed Mrs. Sayles from her son's account. The Sayles family objected, and the matter was brought before the board meeting for Webster Station. On motion, the board "voted that it remain as it is for the present, & Br. [Brother] Browning & Br. Spaulding be a committee to see Francis, and endeavor to get him to pay the debt." The church's control reached beyond town limits. In 1859 residents of East Thompson, Connecticut, a mill town about five miles from Webster, reported seeing Mr. Brown, a member of the Webster Methodist Church, enter a billiard saloon. On his return to Webster, Brown was questioned on his behavior, and when he failed to provide an adequate explanation for his conduct and refused to repent, he was discharged from the church.[41]

Not content only to monitor the behavior of its own parishioners, the Methodist church, in concert with the local Baptist and Congregational churches, tried to translate its dictums directly into

---

church in Webster to conduct investigations of its parishioners. One of the most sensational cases to develop in Webster occurred within the Congregational church. For over a decade Mrs. Stockwell and her husband, a wealthy jeweler, had belonged to the Webster Congregational Church. In the early 1850s she was accused by a fellow resident of adultery. A special church committee was convened to examine the evidence, to extract confessions from the parties concerned, and to make recommendations concerning the disposition of the case. Mrs. Stockwell was called before the assembled congregation and asked to confess her sins in public, to express her sorrow, and to promise to repent. She refused. Dismissed from church, the Stockwells faced a dismal future in Webster. Economically and socially they felt the scorn of their neighbors. In all probability, his business suffered, their children were harassed, and the entire family was cut off from the only society in the township. They soon sold the business and left the village. See United Church of Christ MSS, Congregational Church Parish Records, 1838–1911.

41. See United Church of Christ MSS, Records of Board Meetings for Webster Methodist Episcopal Church, July 1843–July 1856, June 15, 1844, and January 1, 1859, and April 2, 1859; see also Manuscript Schedules, Seventh Census of the United States, 1850.

municipal law and thereby to force everyone in Webster to abide by its concept of right and proper behavior. It succeeded with the temperance issue.

In Webster the assault against liquor coincided with a wider temperance movement gathering momentum throughout the Northeast. Social reformers, including writers on domesticity, led an assault against spirits. In *The Father's Book of Suggestions*, Theodore Dwight, Jr., warned: "Parental example, should be strongly set in opposition to intemperance, every avenue to which should be carefully closed. Incalculable mischief has been done by the toleration of ardent spirits in the house and on the table, even by parents who did not drink it."[42] William Alcott in *The Young Wife* cautioned women to keep their husbands away from the dramshop because they "would go down to a premature grave through the avenues of intemperance and lust, and their kindred vices."[43] Newspaper editors were especially outspoken in their condemnation of liquor. In an article addressed to workingmen, editors of the *Plebeian and Millbury Workingman's Advocate* asked:

> Question—Why do people give liquor to hired work men?
> Answer—For the same reason that an unfeeling man whips a hired horse. The object is to get the most work of them in the least time. It will not do to lay the whip on the backs of free citizens. But they know how to put the whip into your hands, and delude you to goad yourselves to labors beyond your strength. And if you wear out and die, what do they care?[44]

Clergymen, reformed drunkards, and a host of other men and women joined the crusade. Soon one community after another resolved to limit or abolish the sale of intoxicating liquor. Milbury was one of the first communities to restrict its sale. In April 1831 the town meeting resolved to

> view with abhorrence the ravages which Intemperance has committed among us, and that we cordially approve of the efforts now making by our Temperance Societies to suppress its pernicious influence. Resolved, That considering the extensive and injurious effect which

42. Dwight, *Father's Book of Suggestions*, p. 129.
43. William A. Alcott, *The Young Wife, or Duties of Women in the Marriage Relation* (Boston, 1839), p. 89.
44. *Plebeian and Millbury Workingman's Advocate*, May 25, 1831.

ardent spirit has upon the community, by increasing pauperism and crime, that the good of this town does not require the vending of it within the limits of its jurisdiction.[45]

In 1835 the town voted "that the selectmen be requested not to approbate any persons in town to sell wine, beer, or cider."[46]

Although slower to act than other communities, Webster also participated in the temperance crusade. In the early 1830s the Reverend Hubbel Loomis, minister of the Webster Baptist Church, organized Webster's first temperance society, but until his activities were supported by the larger Methodist congregation, his efforts were ineffective.[47]

Officially the Methodist church and its preachers condemned spirits. Church discipline discouraged consumption of hard liquor, and ministers passionately condemned intemperance from their pulpits. "The poor man may be seen drunk upon his last shilling, when perhaps his children are in want of bread," exhorted one itinerant minister. "O, thou invisible spirit of rum, if there is no name thou art known by, let us call thee devil!"[48] But it was not until 1839 that the church took more forceful action. In April the Webster Quarterly Conference "Resolved That in the Judgement of this Quarterly Conference the practice of drinking intoxicating liquors as a beverage is contrary to the rules of the Methodist Episcopal Church."[49] Once the church had taken a stand on this issue, only a few years elapsed before the dictum was translated into law. At a town meeting in 1842, the residents of Webster voted to ban "granting [liquor] Licenses to several persons in this town."[50] Furthermore, the voting procedure adopted at this town meeting broke with convention: the name and the vote of each person par-

45. Ibid., April 6, 1831.

46. *Centennial History of the Town of Millbury, Massachusetts, Including Vital Statistics, 1850–1899* (Millbury, 1915), March 25, 1835, p. 124.

47. Reding, *Historical Discourse Delivered on the Fiftieth Anniversary of the Organization of the Baptist Church*, p. 15.

48. Samuel Allen, *An Oration Delivered on the Anniversary of American Independence, at Blackstone Village, Mendon, Massachusetts, July 4, 1823* (Providence, 1823), p. 13; and *The Doctrines and Discipline of the Methodist Episcopal Church* (New York, 1821), p. 76.

49. United Church of Christ MSS, Records of the Quarterly Meeting Conference, Webster, Mass., April 23, 1839.

50. Webster, Mass., Town Meetings, Minutes: April 11–12, 1842.

ticipating were recorded in the town minutes. The church had a record now of where each of its parishioners stood on the temperance question. Those who voted against the measure risked losing the "Christian sympathy of the church and confidence of the public."[51]

On the temperance issue, the church, the community, and the company stood as one. By the 1830s Samuel Slater and his brother, John, supported the temperance movement and moved to restrict the consumption of alcohol by factory families. In 1837 John Slater of Slatersville cautioned that in the selection of help one should "have special regard to their good morals and the good order of society and they further stipulate that they will not permit the use of ardent spirits by their labouring help."[52] Samuel Slater had also taken action. Daily rations of rum and other spirits allowed workmen were discontinued; in 1832 hands were forbidden to consume spirits on company property, and steps were taken to limit their use in the home.[53] Soon hiring policies came under review, and it appeared that Slater's sons gave preferential consideration to laborers who supported temperance. When he applied for work, Thomas Midgley, for example, stressed his temperance credentials. "As regards my habits," he wrote in 1845, "I belong to the Glenham Temperance Society and has done for this last 2 years and I trust that I can give you satisfaction in this respect. I have a family of six children as self praise is no comendation I think a further discription unnecessary."[54]

Jedediah Tracy, a New York cotton manufacturer, hired only men who disavowed the use of liquor. He ordered that "no liquors could be brought into any workshop under any pretence whatever."[55] Smith Wilkinson, Samuel Slater's kinsman and owner of a Pomfret, Con-

51. United Church of Christ MSS, Records of the Quarterly Meeting Conference, Webster, Mass., April 23, 1839; see also Faler, "Workingmen, Mechanics, and Social Change," pp. 214, 263; and Richard D. Mosier, *Making the American Mind: Social and Moral Ideas in the McGuffey Readers* (New York, 1965 [1947]), p. 127.

52. S. & J. Slater Collection, Lease of J. Slater and others to Moses Buffum and Others, March 1, 1837.

53. Ibid. See also Slater MSS, Sutton Manufacturing Company, vol. 45, October 1832.

54. Slater MSS, Samuel Slater and Sons, vol. 196, Thomas Midgley to Samuel Slater and Sons, July 21, 1845.

55. Letter of Jedediah Tracy, December 27, 1827, cited in White, *Memoir of Samuel Slater*, p. 131.

necticut, factory noted, "In our village there is not a public house or grog-shop."[56] And Welcome Farnum of Waterford, Rhode Island, boasted in 1835, "In our village, with a population of three hundred to four hundred, not an intemperate person lives."[57]

RELIGION AND THE WORKING CLASS

To view religion primarily as a form of social control is to distort its influence on local residents. Religion, especially Methodism, obviously had a strong following among working people. Its discipline, doctrines, and organization appealed to laborers.

Methodism was a version of Calvinism that was well suited to the economic and political realities of Jacksonian America. Although Methodism retained the Calvinist emphasis on right and proper conduct, on long-suffering, and on a simple lifestyle, it rejected Calvinist elitest doctrines. Methodist notions of free will, free grace, equality of all before God, individual responsibility for salvation, and human perfectibility replaced doctrines of prereprobation and infant damnation. As sin was universal, so, too, was salvation. Methodists dropped all artificial obstacles to membership and offered salvation to all who admitted their sins and sought grace.[58] Membership in the congregation and service as lay preachers, stewards, class leaders, and Sabbath school teachers were open to all. During the period shoemakers, wheelwrights, farmers, and merchants served as lay preachers, while operatives, casual laborers, and mechanics served as church stewards. Even the Sabbath school committee was composed of operatives, mechanics, farmers, and mill agents.[59]

An occupational breakdown of the Webster Methodist congregation in 1823, when it was formed, and in 1850, when the church

56. Letter of Smith Wilkinson, Esq., n.d., cited in White, *Memoir of Samuel Slater*, p. 126.

57. Letter of Welcome Farnum, May 23, 1835, cited in White, *Memoir of Samuel Slater*, p. 133.

58. *Doctrines and Discipline of the Methodist Episcopal Church* (1821), pp. 10–21.

59. Daniels, *History of the Town of Oxford*, pp. 366–368, 373, 555; Manuscript Schedules, Sixth Census of the United States, 1840; Slater MSS, Slater and Kimball, vol. 3, Agreements with Help, 1827–1840; Union Mills, vol. 155, Time Book, 1840; United Church of Christ MSS, "Methodist Episcopal Church at the Four Corners."

was firmly established, confirms the wide appeal this sect had among the working classes. Of the original eighteen members of the first Methodist class meeting, organized by John McCausland and Samuel Henderson, both Slater mill employees, eight men worked for the mill as dyers, machinists, pickers, or laborers; three were farmers; and one made scythes. Twenty-five years later the membership included twelve shoemakers, twenty-nine operatives, ten laborers, fourteen skilled craftsmen, eighteen farmers, twelve merchants, and four professional men.[60]

While the equitable character of Methodism accounted partly for the religion's attraction, its doctrines had an even more fundamental appeal to a group of people trying to adjust to and cope with the new factory system and the uncertainties of the wage economy. Religion made comprehensible and bearable the manifold uncertainties, the challenges, and the upsets of everyday life in Webster. Life, after all, was only a testing ground, something that had to be endured. The church advised workers to "run with patience the race which is set before them, denying themselves and taking up their cross daily," to accept the reality of their situation, and to redirect their thoughts and energies toward attaining salvation.[61] The difficulties of life were merely a test of commitment. Methodism partly explained why long-suffering was necessary, but more important, the church provided workers with hope that they might find an eventual release from hardship. For them Methodism was not all darkness, despair, and restraint. Inherent in its discipline was the promise of eternal salvation. Religion was a refuge, and Methodists sought to "find in Christ a hiding place."[62]

Methodist group activities played another, equally important role.

---

60. Daniels, *History of the Town of Oxford*, pp. 366–368; 373, 555. Manuscript Schedules, Sixth Census of the United States, 1840; Seventh Census of the United States, 1850; United Church of Christ MSS, Historical Sketch and Directory of the Methodist Episcopal Church, Webster, Massachusetts (Worcester, 1904), p. 7; "Constitution of the Methodist Episcopal Church Sabbath School, 1830–1840"; "Methodist Episcopal Church at the Four Corners"; Donnell, *Historical Address*, pp. 25, 29–31.

61. *The Doctrines and Discipline of the Methodist Episcopal Church* (New York, 1856), pp. 28–29.

62. *Hymns for Sunday Schools*, p. 65. The remainder of the verse reads: "Lord, may we early seek thy grace, / And find in Christ a hiding place."

Camp meetings, revivals, and Sunday services provided a tempo-rary, immediate release for pent-up passions, anger, and aggres-sion of everyday life.[63] Whenever the Methodists gathered, loud singing, stomping, screams, tears, and hand waving could be heard and seen. The following description of a Slatersville Sunday ser-vice might also describe a meeting at the Webster Methodist Church: "Shoutings were to be heard within and without, while strong men would be falling until some half dozen at a time were lying together upon the floor."[64]

Even more than the regular Sabbath meetings, the revivals and camp meetings were occasions for losing oneself in prayer, singing, exhortations, and a frenzy of physical and emotional activity. Many people were converted at these gatherings. Conversion repre-sented an intensely emotional experience. Benjamin Alcott, while listening to a sermon, found that "the Word reached my heart in such a powerful manner that it shook every joint in my body; tears flowed in abundance, and I cried out for mercy, of which the people took notice, and many others were melted into tears."[65] From Al-cott's description it is evident that these meetings provided workers with an outlet for emotional and physical frustrations. It also is evident that such experiences further reinforced the community spirit within the industrial villages. Emotionally, at least, all those present participated in the experience, and an emotional and spir-itual bond was formed among the participants. The friendship, the understanding, and the sympathy established at revivals extended beyond the four-day meeting and were carried back into the com-munity situation.

Methodism had the potential to act as a radicalizing agent. It drew its communicants from the mass of laboring people, pointed up the harshness and despair in their lives, and provided them with a sense of community and purpose. But such potential was not realized in Webster. While Methodism eliminated barriers to

63. Thompson, *Making of the English Working Class*, pp. 368–369.
64. Buck, *Historical Discourse Delivered at the Semi-Centennial Anniversary of the Sla-tersville Congregational Church*, p. 11.
65. J. M. Buckley, *History of Methodists in the United States* (New York, 1896), pp. 204–206; United Church of Christ MSS, "Methodist Episcopal Church at the Four Corners."

church admission and participation, the Methodist message was not a "democratic theology" or a "frontier faith."[66] Methodists stressed moral and ethical rather than political values. They worked with individuals to attain sanctification and "Christian perfection." Through individual perfection society could be improved.[67] Furthermore, the church's organization could not be considered egalitarian. The church was a structured, hierarchical, centralized organization with a clear line of authority stretching from minister to bishop. The *Doctrines and Discipline of the Methodist Episcopal Church*, hymnals, sermons, and religious tracts approved by the church hierarchy guided the actions and molded the values of the faithful.[68]

In Webster, Slatersville, Wilkinsonville, and the other Slater-style communities, labor militancy and even minor labor disturbances were few. In the decades when workers struck in Dover, Lowell, Chicopee, and Lynn, operatives in Slater's villages went about their activities without appreciable strife. Even when workers in nearby mills clashed openly with manufacturers, Slater's workers did not follow their lead.[69] The structure and atmosphere found there mil-

66. Ahlstrom, *Religious History of the American People*, p. 438.

67. Lois W. Banner, "Religious Benevolence as Social Control: A Critique of an Interpretation," *Journal of American History* 60 (June 1973): 30; Ahlstrom, *Religious History of the American People*, pp. 326–327; Stevens, *History of the Methodist Episcopal Church*, 2:213.

68. Laqueur, *Religion and Respectability*, p. 239; Stevens, *History of the Methodist Episcopal Church*, 2:223.

69. Because Slater's name is closely associated with the Pawtucket textile industry, it is often assumed that Slater and his workers were involved in the strike that took place there in 1824. Their involvement is doubtful. By the 1820s Pawtucket had become a major textile center with eight mills, one of which belonged to Almy, Brown, and Slater. That decade many Pawtucket mill owners (though not Samuel Slater) introduced power looms, and in the spring of 1824 manufacturers cut the wages of recently hired female loom operators. At the same time they extended the workday an hour for all workers by reducing the time allowed for meal breaks. Laborers protested, and female weavers organized a turnout. While many townspeople supported the movement, there is no evidence that Slater's workers, most of them women and children, actively participated in the strike. Almy, Brown, and Slater employed no power-loom weavers in its Pawtucket mill. Neither Almy, Brown, nor Slater signed the statement issued by mill owners after a settlement was reached. Disputes between Slater and his workers were not allowed to explode into the open. See Kulik, "Pawtucket Village and the Strike of 1824," pp. 5–22; Gilbane, "Social History of Samuel Slater's Pawtucket," pp. 266–269; Kulik, "Beginnings of the Industrial Revolution in America," pp. 360–374.

Prude (*Coming of Industrial Order*, pp. 141–144) mentions that a strike took place among hand-loom weavers at Webster in the spring of 1827. He relies on one piece of evidence, a letter from Slater and Howard to John Brown, March 10, 1827, found

itated against the formation of a working-class consciousness. As long as the family and the church remained inviolate and retained their positions within the community and the factory, harmonious labor-management relations continued. In the 1830s, however, the situation began to change. Economic considerations caused many manufacturers, including the sons of Samuel Slater, to introduce cost-cutting programs. Beginning in 1829 and escalating after the death of Samuel Slater in 1835, ownership, management, and labor policies came under scrutiny, and changes were introduced which had an impact on both the labor force and the industrial colonies. By the 1850s the peaceable, compact, traditional societies established by Samuel Slater and his emulators belonged to the past.

---

in the Merino Company Records. The letter says that weavers announced "determination not to weave unless at the old prices." More evidence is needed before "determination" can be termed a strike. And who were these hand-loom weavers and how many participated in this "determination"? Slater employed merchant weavers or subcontractors as early as 1822 to handle much of his weaving. "Merchant Weavers wanted," began an advertisement in the February 6, 1822, edition of the *Massachusetts Spy*. "The subscriber wishes to contract with one or two Merchant Weavers to get from 2 to 4 tons of Yarn woven immediately—for which he will pay a reasonable price in Yarn or Cash. Application to be made at the Factory. Samuel Slater." And in 1827, a few months before the altercation cited by Prude took place: "As you were in the habit of taking weaving from here at the time Mr. Porter was here I thought it would be well to inform you that I have some webs which I can supply your merchants weaving with" (Slater MSS, Slater and Tiffany, vol. 101, Oxford, January 19, 1827). For a discussion of the subcontracting system in south-central New England, see Coleman, "Rhode Island Cotton Manufacturing." The men who did weave for the firm constituted a small proportion of the total number of weavers employed throughout the system.

Part III

THE NEXT GENERATION

[8]

Samuel Slater and Sons:
Business Reorganization

During his lifetime Samuel Slater had acquired a variety of enterprises, from factories to farms. Located in Rhode Island, Connecticut, Massachusetts, and New Hampshire, these enterprises grew
in an uncoordinated fashion. The economic downturn of 1829
caused Slater to reevaluate his policies, and it was at this juncture
that his three sons acquired influence in the business.

To manage the Slater properties, a new family partnership, Samuel Slater and Sons, was organized. The formation of this firm in
1829 signaled the end of an important phase in American industrial development, as commitments to traditional concepts of patriarchalism gave way to competitive capitalism. Confident, competent businessmen, George, John II, and Horatio Nelson Slater,
the three primary partners in the new venture, were willing to break
with tradition, to remove what they believed to be unnecessary im-
pediments to the efficient and profitable operation of the Slater
firms, and to move the family business into the mainstream of
nineteenth-century American industrial development. Under their
direction, the organization and labor policies instituted earlier by
their father came under immediate and thorough review. The
youngest of the three partners, Horatio Nelson Slater, advocated a
complete overhaul of the family business, and he became the chief
architect of the reorganization that followed. Under his guidance,
which lasted for four decades, the firm prospered and became one
of the leading manufacturing companies in the United States. The
credit agency R. G. Dun & Company consistently gave Horatio Sla-

ter high marks for his business acumen: "A very sound concern owned by wealthy stockholders and managed by Scott Mowry one of our most experienced Cotton Manufacturers. No. 1 for all wants," the agency recorded in 1874. "Make 'Pride of the West' the best shirting made in the country. Has been very successfully managed. Owes no debts on its ppty [property]," it reported several years later.[1] The economic success of this family can be gauged partly from the estimates of family wealth made in 1829 by Samuel Slater and later by R. G. Dun & Company. In 1829 Slater assessed his economic worth at between $800,000 and $1 million. The credit agency believed the family was worth considerably less, approximately $500,000 in 1856 and twice that much twenty years later. But in 1876 it had to confess that while "they are currently quoted with over a million dollars . . . no one outside the family can furnish a definite estimate."[2] By all accounts these estimates fell short of the family's real worth. In 1899 the heirs of Horatio Nelson Slater shared over $9 million in stocks, bonds, real estate, and other investments. Their wealth represented only one part of the family fortune, for the heirs of John and George Slater also shared in the prosperity of this business.[3]

The success of Samuel Slater and Sons can be traced to several of the decisions made by the Slater family in the 1830s. Most important, they recognized that the industry had changed. Samuel Slater and Sons manufactured largely fine and fancy cloth, yarn, and thread, and the firm faced stiff competition from British and European mills. Throughout this era, fine goods producers received no protection from the federal government. Although tar-

1. Rhode Island, vol. 10, January 12, 1874, and November 27, 1878, R. G. Dun & Co. Collection, Baker Library, Harvard University School of Business Administration. See also White, *Memoir of Samuel Slater*, pp. 241–242; Slater and Sons, *Slater Mills at Webster*, pp. 6–10; Slater MSS, S. & J. Slater, vol. 15, E. W. Fletcher to John Wright, Providence, March 1838; Sutton Manufacturing Company, vol. 1, 1836–1897.

2. R. G. Dun & Co. Collection, Rhode Island, vol. 9, January 1856; Massachusetts, vol. 97, April 1, 1876. See also White, *Memoir of Samuel Slater*, p. 246.

3. Slater MSS, Estate of H. N. Slater, vol. 35, Commonwealth of Massachusetts Probate Court, Worcester, January 31, 1900; Will, Horatio N. Slater, Jr., July 1891. From an economic and business standpoint, the Slater family was fortunate to have management of the firm pass to a capable member of the family. Elsewhere in New England, this was not the case. As Frederic C. Jaher showed ("Businessman and Gentleman," pp. 17–35), the sons of some Lowell manufacturers avoided the family business.

iffs protected the home market for coarse cloth, legislation did not extend to the fine goods trade. By the 1830s the United States annually imported approximately 3.5 yards of cloth per person.[4] American manufacturers protested. In 1832 the owner of the Arkwright Mills in Coventry, Rhode Island, noted that "the manufacturers of fine goods are still struggling with importations."[5] The Slater brothers concurred; they found it difficult to compete with foreign factory masters, and especially with British producers. In 1835 their Philadelphia agent warned them that

> the real Sea Island thread I would not recommend you to make. It is almost impossible to sell it at a fair price. The article you have herefore sent is thought good enough and to have a better one will operate unfavourably upon it besides, when a better article is wanted buyers almost invariably take the very choicest quality they can select of the many kinds of British spool thread in market.[6]

The Coates thread had an excellent reputation and a lower price tag than the Slater item.[7]

Because they sold their goods at auction, foreign manufacturers were able to undercut many American manufacturers by avoiding the high fees charged by American commission houses and selling goods directly to jobbers and shopkeepers. Domestic producers tried to persuade the federal government to impose a duty on sales at auction. They argued that "from the facilities afforded of effecting speedy sales of all kinds of goods at auction . . . most powerful inducements are constantly offered to foreign merchants and manufacturers to pour the whole of their refuse and surplus productions into our market, to the serious injury of the American trader, and the certain ruin of our manufacturing establishments.[8] The

4. Clark, *History of Manufactures*, 1:248.

5. *McLane Report*, 1:942.

6. Slater MSS, Samuel Slater and Sons, vol. 212, Thomas Remington to Samuel Slater and Sons, May 7, 1836.

7. Ibid., vol. 222, W. Hanson and Bros. to Samuel Slater and Sons, February 10, 1845.

8. "In Favor of Duty on Sales at Auction," *American State Papers, Finance*, Class 4:1037; see also Arthur Harrison Cole, *American Wool Manufacture*, 2 vols. (Cambridge, 1926), 1:216–218; Slater MSS, Samuel Slater and Sons, vol. 232, John Ward to Samuel Slater and Sons, July 15, 1830; vol. 222, Richardson and Trott to Samuel Slater and Sons, February 20, 1838.

auction system, however, did not disappear until demand was regularized and the textile industry matured.[9]

Competition continued unabated. In the 1840s and 1850s, Samuel Slater and Sons again complained about European imports, especially those from Britain. British mills still undercut Slater's prices. The correspondence between Samuel Slater and Sons and its selling agents was revealing. Wells Brothers, the firm's New York agents, wrote in 1847: "The demand for fine goods is limited and prices are very low owing to the free import of foreign goods to compete with them. . . . We learn the late sales have realized a severe loss."[10] From their Philadelphia market the Slaters heard similar news: "The English are sending out large quantities of both brown and bleached which are selling at exceedingly low prices compared with your goods."[11] The complaint was the same several years later: "They have shown us some British, full 36 inches wide, considerably finer quality than yours, which they bought at 7½¢," almost one cent cheaper than the Slater article.[12] In the decade before the Civil War the United States imported approximately 6.5 yards of cotton cloth per person each year. Within a generation, imports of foreign cotton goods had almost doubled.[13]

The Slater family responded pragmatically to this continually increasing competition. Horatio and his brothers recognized that if they were to retain or enlarge their share of the market, they would have to cut costs. Every aspect of their business was reviewed, and they made a number of structural changes in organization. They consolidated some properties, incorporated others, and separated ownership from management. While these organizational forms were known generally throughout the industry and were practiced elsewhere, other changes introduced by the Slater family were more innovative. Samuel Slater and Sons pursued a policy of continuous change and introduced cost accounting, prod-

9. Cole, *American Wool Manufacture*, 1:216.
10. Slater MSS, Samuel Slater and Sons, vol. 211, Wells Bros. to Samuel Slater and Sons, September 21, 1847.
11. Ibid., vol. 210, Whitney Schott and Company to Samuel Slater and Sons, June 5, 1847.
12. Ibid., vol. 211, Whitney Schott and Company to Samuel Slater and Sons, March 3, 1851.
13. Clark, *History of Manufactures*, 1:248.

uct diversification, and the use of brand names; furthermore, it experimented with backward and forward linkages when it tried to integrate into one unit purchasing, production, and marketing. Concerned with long-term growth and stability and determined to maintain their reputation for the manufacture of fine-quality cotton and woolen goods and the integrity of the family name, Horatio and his brothers spent their energy and capital developing the family business.

OWNERSHIP

First the brothers rationalized their holdings by selling some factories and consolidating or enlarging others. Within a decade of their father's death they had sold the Providence Iron Foundry, the Slater and Wardwell store, the Steam Cotton Manufacturing Company, and, in 1846, the Providence Machine Company; in 1848 they sold their interest in the Slatersville property to their cousins.[14] Their interest focused on the Webster holdings, and they attempted to streamline operations there. The three woolen factories were reorganized as the Webster Woolen Company in 1847, and three years later the four cotton mills were consolidated into Union Mills. These factories were enlarged in 1852, 1861, and 1865. Both the Webster Woolen Company and Union Mills were owned exclusively by Samuel Slater and Sons, a closed family partnership.[15]

The partnership, however, was not the only form of ownership adopted by the Slater family. One year after Samuel Slater's death the three brothers broke with family tradition when George Slater, George Wardwell, and Benjamin Hoppin petitioned the state legislature for a charter of incorporation for the Sutton Manufacturing Company. Capitalized at $600,000, this new firm issued 120 shares of stock at $500 per share. The stock of this corporation was never widely dispersed, however, and no new shares were issued.

14. Slater MSS, Introduction and Arrangement of the Slater Collection; Coleman, *Transformation of Rhode Island*, p. 131; Bishop, *History of American Manufactures*, 3:387–388; Bagnall, *Textile Industries of the United States*, pp. 399–401.
15. "Two Hundred Years of Progress," p. 8–9.

Of the stock two-thirds were held by the Slater family, and the rest was circulated privately among close family friends and long-time business associates. Between 1836 and 1860 approximately forty shares of stock passed back and forth among the Hoppin brothers, merchants; George Wardwell, a trader; George Blackburn, a commission agent; and Ezra Fletcher, the clerk of Samuel Slater and Sons.[16]

Although the corporate charter stipulated that a board of five directors and three officers be elected to run the firm, that officials were to meet annually in Providence, that a majority vote determined rules and regulations, and that voting could be by proxy, effective control of the company remained solidly in the hands of the Slater family. Horatio assumed the presidency of the corporation in 1843 and remained in that position throughout the antebellum period. Despite its incorporation, the Sutton Manufacturing Company was little more than an extended partnership.[17]

Under the direction of the Slater family, the company pursued a deliberate expansion program: in 1838 it purchased between 600 and 700 additional spindles and began construction of a new building. Investment lagged during the depression years, but Horatio plowed back into the company all profits earned between 1849 and 1859. During the 1850s no dividends were issued; instead, all profits were reinvested in the construction of buildings and tenements, the purchase of new machinery and equipment, including a steam boiler, and the acquisition of additional real estate in Sutton Township. The company's dividend payments in the 1860s were large.[18] (See Table 10.) All dividends in this period, however, were based on the stock's par value, not on its market value. To the tax commissioner at Providence, Rhode Island, in 1872, Horatio Slater submitted an account of the company, shown in Table 11. This was

16. Slater MSS, Sutton Manufacturing Company, vol. 3, Transfer of Stock, 1837–1880.

17. Ibid., vol. 1, April 8, 1836; Copy of by-laws of Sutton Manufacturing Company; November 8, 1838; Company Minutes: March 3, 1841, February 14, 1842, February 13, 1839, February 21, 1843, February 11, 1846, March 2, 1847, February 12, 1851; see also vol. 45, Administrative Sales, February 20, 1841; H. N. Slater to Tax Commissioner, May 1872, 1880, and 1896.

18. Slater MSS, Sutton Manufacturing Company, vol. 1, Company Minutes: 1836–1897; vol. 45, Administrative Sales, 1818–1899.

Table 10. Dividends paid by Sutton Manufacturing Company, 1839–1868

Date	Dividend as percent of par value
February 1839	8%
March 1841	10
February 1842	6
February 1845	30
February 1846	29
March 1847	15
February 1848	10
February 1860	6
February 1861	10
February 1862	25
August 1862	75
June 1864	30
May 1866	50
November 1866	25
April 1868	20

Source: Slater MSS, Sutton Manufacturing Company, vol. 1, 1836–1897, Company Minutes; and Sutton Manufacturing Company, vol. 45, 1818–1899, Administrative Sales, Memo of Dividends.

Table 11. Value of Sutton Manufacturing Company, 1872

Capital stock	$60,000
Par value per share	500
Market value per share	650
Value of real estate	51,000
Value of machinery	23,000

Source: Slater MSS, Sutton Manufacturing Company, vol. 45, 1818–1899, Administrative Sales, H. N. Slater to Tax Commissioner, Providence, R. I., May 1872.

the first time information on the market value of the stock appeared in company records. Although the Sutton Manufacturing Company prospered, the Slater family did not immediately move to incorporate the Webster property. Until 1865 Webster Woolen Company and Union Mills remained partnerships.[19]

The pace of technological change at the Sutton Manufacturing Company is indicative of that at other Slater firms. Horatio Slater purchased additional machines and constantly replaced or repaired worn-out equipment, but he was slow to introduce new technology. New equipment was often on the market for several

19. "Two Hundred Years of Progress," pp. 8–9.

years before he purchased it, and even then he seldom completely abandoned traditional methods such as the hand mule. Introduced early in the nineteenth century, the hand mule remained in use throughout the antebellum period. When adopted initially by Samuel Slater, it had been an efficient, well-developed piece of equipment, and Horatio Slater saw few reasons to replace it. It was not until 1846 that he started to sell off some of the hand mules and to replace them, presumably, with the self-acting mule. The transition was slow, however, and Samuel Slater and Sons continued to use hand mules; in 1858 Horatio Slater even paid $1,071 for a new one.[20]

Retention of traditional technology could be observed in a second area: motive power. By the 1860s the steam engine could be found in most Slater factories, but its introduction proceeded slowly, and it never replaced water power. Although the Steam Cotton Manufacturing Company ran one of the first factories in southern New England to employ steam power, several decades passed before the Slater family fitted out its other firms with this mode of power. A steam engine was installed at the Sutton Manufacturing Company in 1849, but the firm kept it in reserve; the following year it constructed a new water wheel that served as the primary power source. In all Slater factories, steam engines supplemented rather than supplanted water power.[21]

Throughout the 1830s and 1840s Horatio Slater acquired most of his equipment through the Providence Machine Company, of which he was part owner. After Thomas Hill purchased the company, Samuel Slater and Sons distributed its purchases among sev-

20. Slater MSS, Samuel Slater and Sons, vol. 191, July 1829–July 1830; Sutton Manufacturing Company, vol. 1, Company Minutes: 1849, 1850; see also Jeremy, *Transatlantic Industrial Revolution*, pp. 212–214; Robert R. MacMurray, "Technological Change in the American Cotton Spinning Industry, 1790–1836," Ph.D. dissertation, University of Pennsylvania, 1971, pp. 135–136; Zevin, *Growth of Manufacturing*, pp. 10–41 and 10–42. For purchase of new equipment see Slater MSS, Samuel Slater and Sons, vol. 201, Monroe Osborn and Co. to Samuel Slater and Sons, May 18, 1849; H. Nelson Slater Letters, vol. 33, Calvin Cook to Nelson Slater, April 26, 1841; Samuel Slater and Sons, vol. 197, A. S. Jillsen to Samuel Slater and Sons, April 15, 1846; vol. 190, Central Co. to Storrs, July 31, 1849; Union Mills, vol. 182, April 1, 1858, Paid Bills.

21. Slater MSS, Sutton Manufacturing Company, vol. 1, Company Minutes: 1849, 1850.

eral machine firms, including the Providence Machine Company, Whitin and Sons, William Wheeler, and the Webster Iron Foundry.[22]

MANAGEMENT

Management procedures also were reexamined, and a new supervision scheme was adopted. The brothers established Providence as the permanent headquarters for Samuel Slater and Sons, and while John Slater II managed the office, Horatio Slater supervised operations at Webster, and George Slater set up residence at Wilkinsonville. When important decisions were required, "a general Family interview" was convened in Providence.[23] After John Slater died, in 1838, another reorganization took place. The supervision of the Providence office passed to Horatio, and George moved to Webster. When George died in 1843, responsibility for running Samuel Slater and Sons fell exclusively to Horatio.[24]

To assist him he hired Ezra Fletcher, a former partner in Slater and Wardwell. As company clerk, Fletcher handled all correspondence with cotton factors, retailers, commission agents, manufacturers, and factory agents.[25] Periodically he visited the various mills, inspected the factories, quizzed the agents, and suggested ways to increase output, improve the quality of the yarn and cloth, and, most important, increase efficiency and eliminate waste. To the Webster agent he wrote:

22. Costs charged by the Providence Machine Company conformed to those charged by firms in the area: $3 per spindle for spinning mules, $430 for a spreader, lapper, and doubler. See Slater MSS, Steam Cotton Manufacturing Company, February 11, 1830; January 13, 1832; Union Mills, vol. 179, Paid Bills, 1835–1865; April 2, May 31, June 7, and June 17, 1850; see also Jeremy, *Transatlantic Industrial Revolution*, p. 209.

23. Slater MSS, Samuel Slater and Sons, vol. 235, Samuel Slater to John Slater, April 22, 1831.

24. Ibid., S. & J. Slater, vol. 15, Fletcher to John Wright, March 1838; Slater and Sons, *Slater Mills at Webster*, p. 10; Webster, Mass., Vital Statistics, Deaths, 1843. On the death of George Slater, see Slater MSS, Union Mills, vol. 115, Samuel Slater and Sons to George Bartlett, November 21, 1843.

25. It is not known when the clerk Ezra Fletcher joined the firm. In the Steam Cotton Company records, 1833, there is a reference to him; he may have been an agent or perhaps a bookkeeper for that company; see Slater MSS, Steam Cotton Manufacturing Company, vol. 14, Fletcher to Howe, May 6, 1833.

The writer noticed when last at Webster that a very large portion of the cotton in Picking room was from one lot, vis. W. J. King— Charleston. It is very important when practicable to work all the lots together (that are on hand) in order that the goods may be of uniform quality throughout.[26]

And

> In regard to cotton you appear to dodge the question. Is it a fact that you only weigh a few bales of each lot and if they do not vary much from the bill you say that the "weight holds out"? That is a very loose manner of business. The importance of weighing is as much to detect errors as to find out the average weight.

The agent was ordered "to have all the cotton re-weighed which in plain english means every bale."[27]

Throughout the antebellum period the working relationship between Slater and Fletcher was cordial. In 1843 Fletcher even acquired $10,000 worth of stock in the Sutton Manufacturing Company.[28] But Fletcher's role was clearly that of secretary and adviser: he could not authorize funds or make consignments on his own initiative. Horatio Slater signed personally all bank drafts, set prices for goods and terms of sale, and approved all consignment orders. Repeatedly Fletcher had to apologize to commission agents: "Our Mr. H. N. Slater is not about and we are holding our goods until his return," or "Wishing to consult our Mr. H. N. Slater on the subject is our reason for not answering before."[29] But Fletcher was not the most important agent employed by Samuel Slater and Sons. Horatio placed the actual operation of the factory under specially

26. Ibid., Phoenix Thread Mill, Fletcher to Storrs, August 4, 1847; Union Mills, vol. 187, Fletcher to Storrs, May 9, 1850; vol. 190, Fletcher to Storrs, January 15, 1853.

27. Ibid., vol. 189, Fletcher to Union Mills, March 9, 1854.

28. Ibid., Sutton Manufacturing Company, vol. 3, Transfer of Stock, September 4, 1843.

29. Ibid., Steam Cotton Manufacturing Company, vol. 14, Fletcher to Hallock, April 27, 1833; Samuel Slater and Sons, vol. 203, Fletcher to Samuel Slater and Sons, September 10, 1845; January 13, April 29, September 30, and November 11, 1846; March 1, 1847; Phoenix Thread Mill, Fletcher to Storrs, August 9 and November 6, 1847; April 8, 1848; Union Mills, vol. 188, Fletcher to Union Mills, July 11 and December 5, 1853; vol. 189, Fletcher to Union Mills, March 9, April 18, and December 15, 1854; August 15, 1855; Phoenix Thread Mill, Fletcher to Samuel Slater and Sons, November 4, 1850.

trained, skilled factory agents, following a program initiated by his father in 1829. Over the years he ceded more and more responsibility to these men.[30]

To manage the mills, Slater hired men who possessed a detailed knowledge of textile production but who lacked the capital, the skill, or the incentive to build and operate their own factories. John Clark managed the Steam Cotton Manufacturing Company, Alexander Hodges directed the Sutton Manufacturing Company, and Charles Waite supervised part of the Webster mill complex.[31] Appointed by Samuel Slater, these men continued in their posts after

30. Boston and Providence merchants not only encouraged and financed the construction of the first mills and sold their output through traditional channels, but also influenced the managerial patterns that characterized this industry for over a generation. Frances W. Gregory, *Nathan Appleton, Merchant and Entrepreneur, 1779–1861* (Charlottesville, 1975), pp. 252–265; Gibb, *The Saco-Lowell Shops*, pp. 58–62; McGouldrick, *New England Textiles in the Nineteenth Century*, pp. 21–30; Ware, *Early New England Cotton Manufacture*, pp. 15–38, 61–78; and Chandler, *Visible Hand*, pp. 67–72, among others, have examined the managerial contributions made by these merchant manufacturers. But less is known about the evolution from mercantile-managed factories to those supervised by a group of professional, salaried factory agents. Furthermore, the actual duties and responsibilities of these professional agents remain obscure. Most of the sources available on early factory management apply more to the British than to the American experience.

One of the most thorough accounts of day-to-day factory management and the one cited frequently by American scholars is James Montgomery, "Remarks on the Management and Government of Spinning Factories," in *The Carding and Spinning Master's Assistant, or The Theory and Practice of Cotton Spinning* (Glasgow, 1832), reprinted in *Business History Review* 42 (Summer 1968): 220–226; this account, however, applies to the British factory system. Often cited is Montgomery, *Practical Detail of the Cotton Manufacture of the United States*, pp. 14, 75, 107–109, 155. But machinery, not factory management, was the focus of this study.

Several pamphlets and books were published in the United States on factory management. See Walton Felch, *The Manufacturer's Pocket-Piece*, an interesting and colorful, although not very informative, discussion of factory management. See also Daniel W. Snell, *The Manager's Assistant: Being a Condensed Treastise on the Cotton Manufacture, with Suitable Explanations &c.: to Which are Added, Various Calculations, Tables, Comparisons, &c of Service to the Manufacturer and General Reader* (Hartford, 1850), which was a complete guide for the factory agent. See also Robert H. Baird, *The American Cotton Spinner, and Managers' and Carders' Guide: A Practical Treatise on Cotton Spinning: Giving the Dimensions and Speed of Machinery, Draught and Twist Calculations, Etc.; With Notices of Recent Improvements Together with Rules and Examples for Making Changes in the Size and Numbers of Roving and Yarn* (Boston, 1865).

Experiments in early industrial management were not confined to the textile industry. Several gun-making establishments adopted new organizational schemes during the early nineteenth century. See Paul Uselding, "An Early Chapter in the Evolution of American Industrial Management," in *Business Enterprise and Economic Change*, (Kent O., 1973): 51–84.

31. Slater MSS, Sutton Manufacturing Company, vol. 47, John Slater to A. Hodges, March 16 and April 4, 1831; Wardwell to A. Hodges, July 19, 1831; Steam Cotton

he died. When they were hired, most of them were young, many had begun their careers as machine operators, and all of them were skilled mechanics. Their original contract was for twelve to twenty-four months; they were paid $800 annually in the 1830s and about $1,000 annually the following decade to "see that attention and industry prevails in and about the concern."[32] Agents were cautioned to restrict their attention to the production of cloth and not to allow themselves to be drawn into other affairs because the factory would undoubtedly suffer and "fall short of its proper perfection so long as it has not the undivided care of the superintendant."[33] Specifically, their duties included blending raw cotton for the picking and the carding room, packaging and shipping goods, maintaining accurate records, and recruiting and paying hands—all considered operational functions. The factory agent was a technician who worked under the strict direction of the Slater family.

Mill supervisors were held accountable for the quality of the cloth produced. Horatio usually sent them samples of the type of cloth he wanted and expected them to manufacture it. "Regarding the cloth last sent to you," wrote John Clark of the Steam Mill in 1829:

> The quality I think improved from those you first received still they are not yet what we intend to have them and we have put no mark upon them. We have been greatly troubled in weaving our yarn (spun only upon mules) some webs are very good and others quite imper-

Manufacturing Company, vol. 14, Clark to M. Brown and M. Lewis, February 19 and December 10, 1829; Clark to Tiffany Sayles and Hitchcock, July 1, 1829; Phoenix Mills, Waite to Samuel Slater, July 10, 1834.

32. Ibid., Samuel Slater and Sons, vol. 235, John Slater to A. Shinkwin, July 27, 1829; H. N. Slater Letters, vol. 33, P. Pond to H. Slater, July 31, 1839; Slater and Kimball, vol. 3; Sutton Manufacturing Company, vol. 45, March 9, 1836, Contract Signed by Erastus Walcott; vol. 1 Company Minutes, February 13, 1839; Steam Cotton Manufacturing Company, vol. 15, July 8, 1834. Before hiring agents, the firm required letters of recommendation. One such letter submitted on behalf of Nelson Whitmore read: "Mr. Whitmore is a young man of good character has been acquainted with cotton spinning for a long time is also a good machinist and has had the management of the mill owned by George Weatherhead for several years and I believe to his entire satisfaction. Mr. Weatherhead is still anxious to keep him, but he thinks he can manage a larger concern where he could obtain a larger salary—I see no good reason why Mr. Whitmore would not be a good man for you as a superintendent of your concern" (Slater MSS, H. N. Slater Letters, Leonard Ballore to H. Slater, January 30, 1836).

33. Slater MSS, General Box 1, Samuel Slater and Sons to D. W. Jones, September 26, 1835.

fect. This lot you will find more even than any of the preceeding. I hope in the Spring to arrive at a quality that we may confidently mark and own.[34]

In this case Clark improved the quality of the cloth after he adjusted his machines. But other problems associated with the manufacture of fine and fancy goods could not be solved so easily.

In order to spin fine counts of yarn, the raw cotton used had to be carefully selected and skillfully blended. After weighing and examining each bale of cotton, factory agents selected and mixed various amounts of Sea Island and upland cotton. Tempers flared when the head office supplied agents with short-staple, inferior cotton and still expected them to spin fine yarn. Complaints passed back and forth. In October 1834 Alexander Hodges of Wilkinsonville exploded: "I notice you have carefully avoided saying anything about the main thing in question, that is *sending poor cotton and very small lotts,* I hope Sir you will let this sink deep *in your mind* & when this is thusley connected then I will consider myself as holden responsably for quality and quantity."[35] Rather than manufacture second-rate goods, Hodges threatened to shut down the mill. He set the inferior bales aside for one of the Slater brothers to inspect. Horatio complimented him on this decision. Over the years the Slater name had become synonymous with fine quality, and he was determined to maintain this reputation. Before the cloth was shipped from the factory, Horatio Slater personally inspected it and decided whether or not it was good enough to carry the Slater mark.[36]

In addition to all of his other duties, an agent also spent hours repairing broken machinery, mending reservoir gates, and, in an emergency, sometimes even tending the machines.[37] Moreover, when

34. Ibid., Steam Cotton Manufacturing Company, Letters A, vol. 14, John Clark to M. Brown and M. Lewis, February 19, 1829. Apparently Slater sent to each of his agents samples of the finest cloth woven by both local and British mills and expected them to reproduce it.

35. Ibid., Samuel Slater and Sons, vol. 236, A. Hodges to John Slater, October 3, 1834.

36. Ibid., vol. 212, T. Remington to Samuel Slater and Sons, May 7, 1836; vol. 236, A. Hodges to John Slater, January 27 and May 20, 1834; March 13, 1835. See also Slater and Sons, *Slater Mills at Webster,* p. 10.

37. Slater MSS, Union Mills, vol. 186, Alexander Hodges to Samuel Slater and Sons, November 4, 1837; Sutton Manufacturing Company, vol. 47, A. Hodges to

an agent proved his trustworthiness and competence, he was given additional tasks. Royal Storrs, a twenty-two-year-old Connecticut mechanic employed by Horatio Slater in 1835, demonstrated just how much responsibility an agent could handle. At one time or another, Storrs managed almost every phase of the family's business, supervising not only the Webster factories but also all tenements, boardinghouses, tenant farms, company farms, and company stores in the village. In 1845 he was even put in charge of sales when Slater tried to bypass the commission agents and sell his goods himself. From Webster, Storrs handled all correspondence with jobbers, manufacturers, retailers, and wholesalers; he concluded sales and had the goods shipped to customers.[38]

The agents, however, did not supervise the mills singlehandedly. To assist them, the Providence office hired overseers to manage the carding, spinning, and weaving departments. Each of these operations occupied a separate floor of the factory. Advertisements such as this one for "an Overseer of weaving" in the *Manufacturers' and Farmers' Journal and Providence and Pawtucket Advertiser* outlined the qualities required of overseers: "none need apply but men of skill, industry, and steady habits."[39] In the recruitment of agents and their assistants, family connections now meant less than proven skill and ability. Requested to report to the Providence office, applicants were interrogated by one of the Slater brothers on their previous work experience, their knowledge of machinery, their attitudes toward discipline and toward parental control of children in the factory, and their personal habits. Hired for a trial period of from one to two weeks, overseers were told that "should you give satisfaction we shall in all probability be disposed to want your services at any price which the demand for labour may compel us to pay."[40] Al-

Samuel Slater, September 23, 1835; see also N. B. Gordon, Agent's Diary, November 21 and December 3, 1828; June 5, 1830.

38. Slater MSS, Union Mills, vol. 117, Storrs to R. and D. M. Stebbins, February 28, 1845; Storrs to G. Blackburn and Company, February 13, 1845; vol. 114, Samuel Slater and Sons to Jacob Price Company, November 24, 1840.

39. *Manufacturers' and Farmers' Journal and Providence and Pawtucket Advertiser*, February 20, 1826.

40. Slater MSS, Steam Cotton Manufacturing Company, Letters A, vol. 14, Samuel Slater and Sons to Nelson Swathland, February 11, 1834; see also H. N. Slater Letters, vol. 33, A. Hodges to Horatio Slater, February 28, 1837; Union Mills, vol. 185, Waite to Samuel Slater and Sons, July 6, 1838; Slater and Kimball, vol. 3, Agreements with Help, 1827–1840.

though hired by the central office, overseers and second hands worked under the direction of the factory agent, who was ultimately responsible for the effective operation of the factory.

The recruitment, discipline, and payment of unskilled labor commanded the agent's constant attention. But the labor system he administered differed markedly from the one introduced by Samuel Slater decades before. Changes introduced by the Slater brothers altered the family system of labor and challenged the customary prerogatives accorded householders.

Within a few years of their employment by Samuel Slater and Sons, agents became indispensable to the efficient operation of the business.[41] They became a permanent feature of the new organization, and the best of them were given every encouragement to remain with the firm. Some were even offered a financial stake in the business. The Slater family still believed that owners made the best managers and that only a partner could be trusted to supervise a factory honestly and efficiently. The case of Alexander Hodges was not unusual. When in February 1836 Alexander Hodges decided to leave Samuel Slater and Sons, John Slater II asked him to reconsider. He was prepared to renegotiate his contract and to allow him a $200 increase in his annual salary, and "also allow you to take an interest or possibly we might make a contract by the yard with you."[42] The proposal arrived too late, however, for Hodges

41. Many agents remained with the firm for decades. Royal Storrs, hired by Horatio Slater in 1838 to manage the Webster factories, still held that job in 1860. He was well rewarded for his services. After more than twenty years with the firm he had accumulated $600 worth of real estate and another $23,000 in personal assets. See Manuscript Schedules, Eighth Census of the United States, 1860.

42. Slater MSS, Sutton Manufacturing Company, vol. 47, John Slater to A. Hodges, February 19, 1836. The correspondence between Alexander Hodges and Samuel Slater and Sons is interesting. Hodges left because of a misunderstanding. "You will recollect that during the 4 days you were here and at Webster that you never made me any definite offer to stay," wrote Hodges in February 1836.

You alledged as a reason for doing so, that you must first consult Mr. Hoppin. You will also recollect that Mr. Hoppin said before he left that any arrangement you would make would be satisfactory with him and then I wrote to you on the 17th to know if you had seen Mr. Hoppin—your reply was that *you did not think it necessary* then our mail also brought me news that you had advertised for a man to succeed me on looking into the Journal I found it . . . date of the 9th inst. which was the day before you came to see me and you do not think Mr. Slater that I had good reason to think that my services were not any longer wanted. [Slater MSS, Sutton Manufacturing Company, vol. 47, Alexander Hodges to Samuel Slater and Sons, February 26, 1836]

had already accepted an offer from the Rockingham Manufacturing Company to manage its factory. Ezra Fletcher, on the other hand, accepted such a Slater offer.[43]

ACCOUNTING

By the 1830s the first step had been taken to separate entrepreneurial functions from operating functions. Yet the Slater brothers required an effective check on the efficiency and honesty of their agents. John's death left the firm shorthanded; both Horatio and George increasingly were drawn away from the factory and towns and could no longer exercise direct, constant surveillance over factory agents. Some method had to be devised to monitor their activities, to ensure their honesty and diligence. Furthermore, as competition increased in the 1840s and the costs of production continued to climb, they also wanted to devise some method whereby they could determine with accuracy the actual costs incurred in the manufacture of goods.

When the Slater brothers assumed control of the business, the accounting methods used were simple and straightforward. In the 1830s agents spent hours with their bookkeepers recording all expenses involved in the manufacture of cloth, yarn, and twine. Their

---

In reply John Slater wrote:

I do not perceive what should cause you to use the following . . . 'I now take it for granted that my services are not any longer wanted.' If I recollect right I named to you that I did not think it necessary to see Mr. H. [Hoppin] in as much as you stated that you must give an ans. in Boston on Monday (the 15th inst.) I did not even so much as dream that you meant next Monday the 22 inst. but on receiving your favor of the 17th I thought I must be mistaken or mave misunderstood you, I therefore after writing you I called to Mr. H. who also seemed to be impressed that it was last Mon. you had reference to but when we both come to find ourselves mistaken we concluded we might afford to allow you about $200 advance from what you had been having & also allow you to take an interest or possibly we might make a contract by the yard with you but as you *must* go into Boston on *Monday* and conclude a contract with the other company. . . . to make arrangements with you perhaps you may discover some way to *postpone* it until the Monday after and in the interim take an opportunity to reach an understanding. If not I wish you to endeaver to be absent as little as possible and not think of leaving before your contract is up (on the 1st. of April) as a contract is a contract as regards time if not as respects price. [Slater MSS, Sutton Manufacturing Company, vol. 47, John Slater to Alexander Hodges, February 19, 1836]

43. Ibid., A. Hodges to Samuel Slater and Sons, March 11, 1836.

short reports, submitted monthly to the central office, noted the number of yards manufactured and the problems, if any, encountered at the mills. One such memorandum read: "this is to inform you how we are getting along with the manufacturing business at this place & I have had a tolerable good supply of all kinds of help for the last four weeks got off 42,000 yards of cloth—the prospect for the four weeks to come is not as good as the past."[44] Within a decade, however, this simple procedure was replaced by a more elaborate one. As early as 1839–40 an elementary form of cost accounting was introduced.[45]

The accounts kept by the Sutton Manufacturing Company agent typified the new system. At this firm the agent was required to keep a variety of accounts, from wage and expense books to a factory ledger titled "Expenses of Sutton Manufacturing Company at Sutton Mill." From the items recorded in the factory ledger, the agent calculated monthly the cost per yard for labor and so-called incidental expenses and also the entire cost per yard for cloth.[46]

The factory ledger, which was important to the cost accounting system introduced by the Slaters, was divided into months. Each month was then separated into accounts for the various departments: carding room, stretching and spinning room, mule room, dressing room, and weaving and cloth room. Entries for material used and certain overhead expenses incurred in the operation of each department, such as leather, tacks, rollers, oil, coal, miscellaneous labor costs, and machine repairs, were made daily. The agent labeled these costs "incidental" expenses. At the end of the month direct costs for labor, raw materials, and freight were transferred from separate control accounts and were charged to the individual departments. Costs for postage, real estate, transportation, supervision, bookkeeping, interest, insurance, and costs incurred in guarding the mill were transferred to the monthly account and

44. Ibid., Samuel Slater and Sons, vol. 236, A. Hodges to John Slater, August 18, 1834; Sutton Manufacturing Company, vol. 47, Hodges to Samuel Slater, August 31, 1835, and March 14, 1836; Slater and Tiffany, vol. 101, August 31, 1836; Samuel Slater and Sons, vol. 236, A. Hodges to John Slater, February 19, 1834. See also Chandler, *Visible Hand*, p. 69.

45. Slater MSS, Sutton Manufacturing Company, vols. 31 and 32, Expenses, 1839–1856; vol. 33, Expenses, 1872–1887.

46. Ibid., vols. 31 and 32, Expenses, 1839–1856; vol. 33, Expenses, 1872–1887.

were recorded separately. All accounts were then closed and a summary was written. This summary included the various departmental costs and also the other costs incurred that month. From this information the costs involved in the manufacture of one yard of cloth were calculated. This procedure was quite simple: once again the monthly account was broken down into departmental accounts, and for each department the costs for stock, labor, and incidental expenses were given. Then the cost per yard was computed. In the carding room two sets of figures were reported: the cost per yard for stock, labor, and incidental expenses, and the cost per yard for labor and incidental expenses. In the remaining departments, only the second cost was provided. A figure was also calculated for various other expenses. The account was closed, and the final unit costs were determined. In July 1840, for example, the cost per yard for cloth manufactured by the Sutton Manufacturing Company was 4.433 cents; of this amount, labor and incidental expenses totaled 2.449 cents.[47] Once specific information was made available, steps could be instituted to trim those expenses. This cost accounting system was to have important ramifications for both Slater and his workers. Labor proved to be the most costly item in the manufacturing process, and if Slater wanted to lower costs, labor was the obvious target.

This cost accounting system was one of the most sophisticated forms of accounting employed by early-nineteenth-century American businessmen. In the 1840s, when almost all other manufacturers, including those who patterned their operations after the Lowell system, still employed double-entry bookkeeping, the Slater family devised an accounting scheme that allowed it to pinpoint expenses. It appears that other firms did not adopt a similar accounting system until the 1850s.[48]

47. Ibid.
48. H. Thomas Johnson ("Early Cost Accounting for Internal Management Control: Lyman Mills in the 1850s," *Business History Review* 46 [Winter 1972]: 466–474) found that a fully integrated double-entry cost accounting system operated in the Lyman factories in the 1850s. He suggested that this mill was not unusual, that perhaps other mills employed some form of cost accounting, as, indeed, the Slater firm did. See also Chandler, *Visible Hand*, pp. 70–71. While the Slater accounting system represented an important break with traditional mercantile accounting methods, it was not without its drawbacks. The agent combined fixed and variable costs; furthermore, factory costs were not separated from administration costs. See Slater MSS, Sutton Manufacturing Company, vols. 31–33, Expenses, 1839–1887.

MARKETING AND PURCHASING PROCEDURES

Innovations in business administration extended beyond the production process itself, as Horatio Nelson Slater sought control over the marketing of his goods and the supply of raw materials used in his mills. Such control required him to eliminate commission agents who handled these responsibilities, or at the least to limit their influence.

Like most New England manufacturers in the 1820s and 1830s, Samuel Slater had consigned his goods to specialized selling houses in the major commercial centers of Boston, New York, Philadelphia, Baltimore, and Charleston. But unlike the Merrimack, Hamilton, and Pepperell firms, which channeled goods to one house, Samuel Slater tried to prevent large amounts of cloth and other goods from falling into the hands of a single dealer. In the Philadelphia market, for example, he employed a minimum of seven commission agents, including R. Fisher; Brown, Hanson and Company; Eli Rising; Brown Brothers; Hall and Company; Thomas Remington; and Richardson and Trott. Although an agent might charge that "if several houses have them here they will get the price down as a matter of course," Slater ignored the complaint.[49]

Initially Slater allowed his agents to "sell these at a price your own judgment may warrant, without any special recommendation."[50] But because agents failed to provide him with detailed in-

49. Slater MSS, Samuel Slater and Sons, vol. 210, Hunt Brothers to Samuel Slater and Sons, September 10 and September 21, 1839. Rather than consign his goods to a variety of houses, Francis Cabot Lowell consigned all goods to B. C. Ward, a Boston agent who handled Lowell goods for over a decade. But in 1828 several large stockholders in the major Lowell-style firms organized their own selling house, J. W. Paige and Company. This firm exercised considerable authority over its suppliers, not only finding customers, establishing their credit rating, and transacting all business with them (including arrangements for transportation, storage charges, and other facilities), but also advising manufacturers on fashion trends and market prices, and even extending loans and other financial assistance to them. See Fred M. Jones, "Middlemen in Domestic Trade, 1800–1860," in *Readings in the History of American Marketing: Settlement to Civil War* ed. Stanley J. Shapiro and Alton F. Doody (Homewood, Ill., 1968), p. 281; John S. Ewing and Nancy P. Norton, *Broadlooms and Businessmen: A History of the Bigelow-Stanford Carpet Company* (Cambridge, Mass., 1955), pp. 27–29; Evelyn H. Knowlton, "Textile Selling Agents," in *Readings in the History of American Marketing*, ed. Shapiro and Doody, pp. 379–380; Paul T. Cherington, *Wool Industry: Commercial Problems of the American Woolen and Worsted Manufacture* (Chicago, 1916), pp. 117–119; Ware, *Early New England Cotton Manufacture*, pp. 178–182; *Niles Weekly Register*, March 1, 1834.

50. Slater MSS, Steam Cotton Manufacturing Company, vol. 14, Clark to M. Brown and M. Lewis, February 19 and December 10, 1829.

formation on prices prevailing in their markets, and because he thought the selling houses were not doing enough to advance the sale of his goods, he established a new policy. To Tiffany Anderson, a New York agent, he wrote in 1831:

> Not one among you has named any price which might be expected for other shipments. Now with us who make goods the price is an important consideration. These goods have been quite dull and heavy since the demand which formerly existed for export has ceased.— and for the last year the prices have been quite unsatisfactory; so much so that I feel like availing myself of the earliest opportunity of putting them at the following price (the lowest at which they should stand).[51]

Thereafter Samuel Slater set prices and the terms of credit.

When his sons tried to pursue similar policies, they found it impossible to do so. Because the Slater family insisted on setting and maintaining prices, and because these prices often were higher than those prevailing in the market, they frequently had to guarantee all sales. Under this arrangement, the Slater brothers set prices and established the terms of sale. Commission agents accepted Slater goods on consignment, found customers for their goods, negotiated contracts, and then turned those contracts over to Samuel Slater and Sons for payment collection. If the customer proved to be a poor risk and defaulted on his note, the company, not the commission agent, suffered. On this type of sale, the commission agent received a fee of between 2.5 and 4 percent for his service. If, on the other hand, the agent assumed all risks and guaranteed sales, he earned 6 percent.[52] In either case, Samuel Slater and Sons paid higher commission rates than many competitors. J. W. Paige charged the Waltham, Merrimack, Hamilton, and Appleton companies between 1 and 1.25 percent for services.[53]

51. Ibid., Steam Cotton Manufacturing Company, vol. 14, Samuel Slater to Tiffany Anderson, March 17, 1831.
52. *McLane Report*, 1:576–577; Slater MSS, Samuel Slater and Sons, vol. 210, Underhill and Company to Samuel Slater and Sons, July 25, 1845, and May 14, 1846; Hunt Brothers to Samuel Slater and Sons, July 25, 1846; vol. 216, Tiffany Anderson and Company to Samuel Slater and Sons, March 8, 1834; vol. 232, B. Newboldt to Samuel Slater and Sons, January 1, 1838; Union Mills, vol. 114, Samuel Slater and Sons to T. Remington, February 6, 1841; Samuel Slater and Sons, vol. 236, R. Fisher to John Slater 2d, April 28, 1830.
53. Ware, *Early New England Cotton Manufacture*, pp. 178–179; Appleton, *Introduction of the Power Loom*, p. 12.

In most cases the Slater company, not the commission agent, set the credit terms and organized transportation. Credit terms varied from one year to the next. In the 1830s six to eight months' credit was common. By the 1840s, however, four months' credit, with liberal discounts for cash, became common practice, although the rates fluctuated. Union Mill sheeting, one of the company's fastest-selling and cheapest products, was consigned for 8¾ cents per yard, ninety days' credit, in March 1841; 8 cents per yard, eight months' credit, in October 1842; 4½ cents per yard, four months' credit, in September 1843; 7¼ cents per yard cash in August 1843; 7¼ cents per yard, four months' credit, in March 1845; 7½ cents per yard, sixty days' credit, in July 1845; 8¼ cents per yard, four months' credit, in October 1845; 7½ cents per yard, four months' credit, in July 1846, and 7 cents per yard, four months' credit, in September 1846.[54] Changes were evident in the discounts given for cash transactions as well. In October 1840, for example, cash discounts of up to 7 percent were allowed on this type of cloth, but by March 1841 discounts had dropped to only 2 percent. In the fifties, six months was again the accepted credit term allowed agents—no barter.[55] For the most part, agents paid the cost of transportation from the Webster or the Worcester railroad depot or, if the shipment was to go by sea, from Providence to their place of business.[56]

The fees paid to middlemen, combined with competition from foreign manufacturers and the high cost of production, threat-

54. Slater MSS, Union Mills, vol. 114, Samuel Slater and Sons to A. Collins and Son, March 27, 1841; Samuel Slater and Sons to Wells and Spring, October 8, 1842; ibid., vol. 116, Samuel Slater and Sons to Stoddard and Lathrop, August 2 and Samuel Slater and Sons to T. Underhill, September 7, 1843; vol. 190, H. Slater to Storrs, March 28, 1845; Samuel Slater and Sons, vol. 210, Underhill and Company to Samuel Slater and Sons, July 3, July 25, and October 20, 1845; Hunt Brothers to Samuel Slater and Sons, July 25 and September 15, 1846; Underhill and Company to Samuel Slater and Sons, September 3, 1846.
55. Ibid., Union Mills, vol. 114, Samuel Slater and Sons to Charles Hall, October 7, 1840; Samuel Slater and Sons to A. Collins and Sons, March 27, 1841; vol. 188, Fletcher to Union Mills, November 5, 1852; vol. 189, Samuel Slater and Sons to Union Mills, May 7, 1855.
56. Ibid., vol. 114, Samuel Slater and Sons to Stoddard and Lathrop, January 20, 1840; vol. 116, Samuel Slater and Sons to T. Underhill, September 7, 1843; vol. 210, Underhill and Company to Samuel Slater and Sons, September 3, 1846; Union Mills, vol. 188, Fletcher to Union Mills, November 5, 1852. These commission agents provided Slater with the usual services; they solicited customers, advised on style and fashion, reported on the competition, and offered financial assistance. As a rule, however, the firm did not borrow money from agents or encourage them to invest directly in any Slater project.

ened to price Slater goods out of the market. To combat these problems, the Slater brothers devised a number of schemes. In the mid-1830s they designed a stamp to be placed on their manufactured goods. Although Samuel Slater and Sons did not pioneer the use of brand names, it realized early the attraction its name held for the public. One of its most successful stamps was "Triumph of American Manufacturers." Capitalizing on the patriotism engendered by the Mexican War, Horatio Slater was able to advertise successfully and to market his goods under the Triumph ticket. For several years his agents could not stock enough of the Triumph cloth to meet demand.[57] Underhill and Company implored Slater: "March 20 begins our order (given last fall) for unknown's to be stamped 'Triumph of American Manufacturers' as agreed upon then, and I hope you will not forget it, as we expect to make a great display with the goods from the 'Sons of the first manufacturer in the United States.'"[58]

A more serious attempt not only to meet competition but to lower prices occurred in 1845 and again in 1849, when Horatio Slater abandoned consignment sales altogether and tried to sell his goods directly to agents, jobbers, merchants, shopkeepers, and manufacturers. Commission agents were informed "that we do not now have any cotton goods to consign as we are selling them at the mill as fast as made."[59] Agents who wished to carry Slater goods had to buy them outright. To former commission agent George Blackburn and Company, Horatio Slater wrote: "As we do not wish to consign any of our first quality shirtings we propose to sell you as many as you want at 8 cents less 10 percent for cash delivered at the Depot here, which will give you a large commission than I would

57. Ibid., Union Mills, vol. 185, Waite to Samuel Slater and Sons, August 3, 1836; Samuel Slater and Sons, vol. 210, Hunt Brothers to Samuel Slater and Sons, August 22, 1836; vol. 212, Thomas Remington to Samuel Slater and Sons, May 7, 1836; vol. 210, Whitney, Schott, and Company to Samuel Slater and Sons, February 8, 1848; vol. 217, Hacker Leat and Company to Samuel Slater and Sons, March 28, 1848; see also Joseph C. Robert, "An Early Use of Brands and Trade names," in *Readings in the History of American Marketing,* ed. Shapiro and Doody, p. 409.

58. Slater MSS, Samuel Slater and Sons, vol. 210, Underhill and Company to Samuel Slater and Sons, March 17, 1846.

59. Ibid., Samuel Slater and Sons, vol. 117, Samuel Slater and Sons to R. & D. M. Stebbins, February 28, 1845. It appears that Slater may have tried this scheme in 1840 as well. See Union Mills, vol. 114, Samuel Slater and Sons to Jacob Price Company, November 24, 1840.

to occur . . . in consignment."[60] But Slater's optimism faded in July when trade declined, and he again had to return to the consignment system. Another attempt to sell goods directly from the factory occurred four years later, but within a year Slater once again returned to traditional outlets.[61] Although in the following decade Slater continued to flirt with this idea, he did not succeed in bypassing these middlemen until 1866, when he opened a warehouse in New York and sold goods directly to wholesalers and retailers. It is believed that "S. Slater and Sons, Inc. are the oldest commission house in the country . . . to sell their own goods."[62]

In the purchase of cotton, however, the Slater family was able to bypass commission agents and thus save money. Initially Samuel Slater and Sons ordered Sea Island and the finest grades of upland cotton through commission agents in Charleston, Mobile, and New York. In the 1830s agents scoured the Charleston market for quality cotton on behalf of Joseph Leland and Brothers, who served the Slaters in that region. But the relationship between the manufacturers and the agents seldom ran smoothly, for the prices set by the Slaters rarely coincided with those prevailing in the local market. In 1834, for example, Joseph Leland and Brothers advised:

We regret to say your order for cotton cannot be executed. Our market having improved very considerable and is not in active demand at our quotations. . . . In Sea Island there has been a considerable business done also, there is one fine parcel which we could purchase at 26 cents which we think would suit you, but none that would answer to be had at your limit of 25 cents. We shall keep an eye on what may come in and if we meet with such a parcel as will answer at your limits shall purchase and ship giving you timely notice of our intentions to do so. The Upland you will see cannot be had now and the prospect is we shall not be able to get such as you want at limits at least for some time to come.[63]

60. Ibid., Samuel Slater and Sons, vol. 117, Samuel Slater and Sons to G. Blackburn and Company, February 13, 1845.
61. Ibid., Samuel Slater and Sons, vol. 190, Caufield to Storrs, October 26, 1849.
62. Ibid., vol. 198, Mills and Co. to Samuel Slater and Sons, August 21, 1850; H. Low to Samuel Slater and Sons, February 9, 1852; Slater and Sons, *Slater Mills at Webster*, p. 33.
63. Slater MSS, Samuel Slater and Sons, vol. 205, Joseph Leland and Brothers to Samuel Slater, March 29 and April 23, 1834; June 25, 1838.

When the commission agents finally secured and packaged a shipment, other complications could develop. On inspection of the bales, factory agents sometimes found that cotton was packed deceptively. In some cases inferior cotton and dirt were mixed with the finer grades of cotton, while in other cases the staple was injured because the cotton had been packed while still wet. Slater would then complain, and a grievance committee of three would be appointed to inspect the goods and settle the account.[64]

Unlike other manufacturers, the Slater family did not allow one agent to monopolize the trade. In the Mobile market it traded with Judson Hoppin and C. W. Ogden, while in the New York market it bought cotton from A. J. Kitchum and N. Talcott and Sons. Closer to the mills, the northern houses provided the same services as their southern counterparts, but the quality of the cotton they supplied was often poor. In a letter dated May 10, 1834, A. J. Kitchum apologized: "We regret to learn that the 50 bales cotton does not prove as good as you wish. It is true, the parcel does not equal the prime qualities of 'other years.'"[65] In this as in so many other cases, Slater returned the cotton to the dealer labeled "unacceptable."[66]

Because the Slaters failed to receive sufficient supplies from regular agents, and because they wished to avoid the commission house fee, they sought to establish an alternate method of procuring raw material. They hired itinerant peddlers to attend sales meetings and auctions in communities throughout the South and to purchase cotton on their account. Charles Rogers of Mobile was one of these agents. In 1834 he attended sales meetings in Columbus, Georgia, and Apalachicola, Florida, where he found that the excellent staple and silky quality of the cotton rivaled that of the finest New Orleans product. It was purchased and shipped directly to the Slater firm.[67] By employing such men as Rogers, Samuel Slater and Sons cut costs considerably and pioneered an alternate

64. Ibid., S. & J. Slater, vol. 15, Shubael Hutchins, Salena Manton, and Rila Scott to Samuel Slater and Sons, November 21, 1834.
65. Ibid., Samuel Slater and Sons, A. J. Kitchum to Samuel Slater and Sons, May 10, 1834; N. Talcott and Sons to Samuel Slater and Sons, December 31, 1835; Corey Taber to Samuel Slater and Sons, February 15, 1837.
66. Ibid., vol. 236, Samuel Slater and Sons to John Slater, Esq., March 13, 1835.
67. Ibid., vol. 205, Charles Rogers to Samuel Slater, December 8, 1834, and November 21, 1838.

method of cotton procurement. Following the Civil War, peddlers became prominent figures at cotton sales, and many northern factory masters turned to them for their supplies.

The Slater brothers' reassessment of ownership, technology, management, and accounting, marketing, and purchasing procedures reduced costs somewhat and increased efficiency, but prices were still too high. To compete successfully against foreign importers and domestic manufacturers the Slaters had to review other expenses. Approximately half of the cost involved in the manufacture of a yard of cloth was accounted for by labor and allied expenses. This area represented the largest potential for cost reductions. The death of Samuel Slater gave his sons the opportunity, the increased competition evident in the industry gave them the excuse, and the cost accounting system provided them with the information necessary to institute wide-ranging changes in the traditional system of labor. The family system of labor was submitted to close scrutiny and was found wanting. It appeared to be an expensive and inefficient use of human resources.

[9]

Management vs. Labor

At the turn of the nineteenth century the family system of labor had served the needs of both employers and employees. Economic and social circumstances peculiar to the era had brought about this conjunction of interests. Yet this system lasted barely one generation. With the growth of the market economy, and with increasing competition from foreign producers, the interests of workers and managers diverged. Management attempted to salvage its own economic position in part at the expense of labor by dismantling rights long accorded householders and workers.

The dissolution of the traditional family system of labor occurred over several decades. Efforts to change customary worker prerogatives proceeded slowly and in a piecemeal fashion. Although labor and management disagreed on some issues, patriarchy initially remained the context in which disputes were decided. Nevertheless, the unmistakable force of events moved against familial influence. Slowly the firm chipped away at traditional rights, advancing and retreating on issues as events and the economic situation dictated. By the 1850s little remained of the labor system established by Samuel Slater at the turn of the century. Authority over hiring, firing, and disciplining workers, allocating jobs, and setting working conditions rested firmly in Horatio's hands now. The householder was stripped of his influence within the factory, and Slater dealt separately with individual workers.

Between the mid-1820s and 1850, the decline in the family system of labor and in the influence exercised by householders could be observed in two areas. First, the composition of the work force changed. Single, itinerant hands found employment in Slater's industrial communities. Second, work rules and procedures within the factory were altered as management assumed all rights to recruit and discipline workers. These two issues were interwoven throughout the period; changes in the composition of the work force and alterations on the factory floor proceeded in tandem.

DECLINE OF FAMILY LABOR

Initially through advertisements placed in urban and rural newspapers, Samuel Slater and many other manufacturers south of Boston had recruited sufficient hands to keep their machines operating. Beginning in the 1820s, however, this labor reservoir began to dry up. As more manufacturers entered the industry, competition for labor became keen. Those who could not secure enough help through traditional channels began to adopt a more drastic and more direct approach to the labor problem: they pirated hands from their neighbors. Cooperation gave way to open competition, and by the early 1830s poaching was common throughout southern Massachusetts.

Slater's factory towns were prime targets. In April 1834 Alexander Hodges of Wilkinsonville complained to the head office:

> I have not been able to make a great quantity for the last two weeks—one week there were only 30 looms in operation. Last week, 50, this week I expect to run 60—the great difficulty is this—the Newton Falls folks send through these parts every week after weavers they offer 44 cents per cut—which has caused some of my weavers to leave—then there are . . . Kimball's from Holdon here every few days after all kinds of help.[1]

Throughout the early 1830s Hodges seldom had enough weavers and mule spinners to operate his machines. In February 1834 he advised the Providence office that "most of the mills have commenced hiring their help—help will be in great demand in this

---

1. Slater MSS, Samuel Slater and Sons, vol. 236, A. Hodges to John Slater, April 14, 1834.

vicinity this spring."² The situation deteriorated. "None of my neighbors have as yet learned the prices of help," wrote Hodges two months later, and although "I have seen them on the subject they say they will cut down before long—that is after they get filled up &c &c—I think I shall be able to fill up before long so as to run 70 mules."³ As late as August of that year, Hodges still had not hired enough hands to operate at full capacity. "I am already short of weavers & some more have given their notice to leave to go to other places where they can do better as they say."⁴ When labor shortages occurred, Hodges requested and usually received workers from other Slater factories. Because he owned mills throughout New England, Slater could shift workers from one area to another. When the widowed Priscilla McKay, for instance, applied in 1831 to John Clark at the Steam Cotton Manufacturing Company for jobs for four of her children, she was told that Clark did not require her family, but Alexander Hodges at the Wilkinsonville mill could use them.⁵

Relocation of some workers did not always solve the problem. Manufacturers had to mechanize, increase the productivity of their employees, or find another source of labor. Initially most manufacturers chose the last solution, falling back on a traditional source for additional help. For a generation children had formed the basis of the factory labor force, and boys and girls seemed the obvious choice. Apprentices, of course, could be introduced again, or perhaps workhouse or pauper children could be hired on contract. Neither method of child employment, however, allowed additional hands to be integrated easily into the existing order. Industrial villages and factory operations had been designed to meet the requirements of family labor. Housing, wage settlement patterns, recruitment policies, disciplinary modes might have to be altered to

2. Ibid., Sutton Manufacturing Company, vol. 47, Alexander Hodges to Samuel Slater, February 2, 1834.
3. Ibid., Samuel Slater and Sons, vol. 236, A. Hodges to John Slater, April 14, 1834.
4. Ibid., August 18, 1834.
5. Ibid., Steam Cotton Manufacturing Company, vol. 11, Providence Journals, October 18, 1831; Sutton Manufacturing Company, vol. 47, John Slater to A. Hodges, April 4, 1831; Wardwell to Hodges, July 19, 1831; John Slater to Hodges, March 16, 1831; Samuel Slater and Sons, vol. 236, Hodges to John Slater, April 14, 1834.

accommodate these child laborers. Family government might be challenged, and in the 1820s and early 1830s, manufacturers were not ready to anger householders by introducing novel forms of labor. Furthermore, if past experience was a guide, apprentices could prove hostile and expensive. Rather than risk quarreling with householders or squandering capital on questionable ventures, manufacturers urged parents to recruit and board additional laborers themselves. The entire problem of labor recruitment, as well as part of the disciplinary function, now fell on the householders' shoulders. New England newspapers ran want ads for "two or three families with help."[6] Householders assumed part of the responsibility for recruiting hands for the textile industry.[7] It might seem that this development would serve to strengthen the family system of labor, securing it more firmly than ever within the textile towns.

At first householders probably recruited their help from among children who were indigent, abandoned, or otherwise neglected and under the temporary care of poor-law officials. Certainly a pool of these children existed, for although rural communities built almshouses and operated pauper farms for the long-term care of indigent or elderly people, few facilities existed for children. Children aged four and upwards were bound over to those willing to provide them with food, shelter, and clothing.[8] Massachusetts law legally recognized the town's prerogatives in such cases and urged overseers to "bind out the minor children of any poor person who has become chargeable to their town."[9] This was the case in Oxford, where one young widow, Mrs. Webster, obtained the guardianship of Mary Kingsbury, daughter of Rufus Kingsbury, a local troublemaker and ne'er-do-well, and placed the child, along with

6. *Massachusetts Spy* (Worcester), June 8, 1831, quoted in Ware, *Early New England Cotton Manufacture*, p. 212.

7. Tamara K. Hareven, "Family Time and Industrial Time: Family and Work in a Planned Corporation Town, 1900–1924," *Journal of Urban History* 1 (May 1975):269–270. In her study of Manchester cotton textile workers, Hareven stresses the role played by the family as a recruiting agent. The scheme she describes has roots in the antebellum period.

8. *Niles Weekly Register*, November 11, 1826; Emerson, *History of the Town of Douglas*, pp. 55–56; see also Klebaner, "Pauper Auctions," pp. 195–210.

9. Massachusetts, Revised Statutes (1836), Chapter 46, secs. 22, 372; see also Webster, Mass., Town Meetings, Minutes: March 11, 1839; March 2, 1840; March 7, 1842.

her own two daughters, in the local Slater mill, where they remained for the next eight years.[10] At this mill one prospective employee placed six children, four of his own and two others; another agreed to supply three or four extra hands besides himself and his four children; and a third collected wages for seven children, three of his own and four others.[11] Through this scheme, Slater was able to increase the ratio of factory workers to farm laborers, teamsters, and other comparatively well-paid hands and to solve temporarily the labor shortage problem.

Under changing economic circumstances, the family proved flexible. This forced enlargement of the family unit, however, had enormous consequences for all concerned. Increasingly the family became a work unit; its members' ties to one another were economic ones. Moral or religious considerations and obligations were less important than economic reasons for adopting or taking in charges, and the strength of these new bonds was open to question. How long would these youngsters remain with the family and contribute to its economy? Would they submit easily to family government? Although initially this new form of labor recruitment represented an adjustment between labor and management, in time the practice came to represent a step in the undermining of the family's position in the factory and the community. Samuel Slater had set a precedent. Itinerant hands could be introduced into the factories. At this junction they were still under family government, but they would not long remain there.

This situation represented a crisis in the relationship between factory master and householder, and it was resolved at first within the patriarchal ideology that united the two forces. Each side conceded certain issues: labor recognized that additional hands were required; management recognized the right of the householder to control the new workers. Through negotiation, not protest, the disagreement was settled. Additional hands were employed and were placed under family government.

But this compromise did not completely solve labor problems, and in the following decade Samuel Slater and Sons began to re-

10. Slater MSS, Slater and Kimball, vol. 3, Agreements with Help, 1827–1840. Single-parent families comprised a significant proportion of the textile labor force.
11. Ibid.

cruit additional workers. Targets for this recruiting drive were primarily young women. The firm appeared able to convince householders that it had legitimate reasons for the introduction of female
hands and that these workers supplemented, not supplanted, family labor.

These itinerant workers were hired to fill places in the weaving
room. Conversion to the power loom, begun hesitantly by Samuel
Slater in the 1820s, was completed the following decade. By 1840
power-loom weavers comprised approximately 25 percent of the
labor force.[12] Because of the precedent set by Francis Cabot Lowell, and because weaving was thought too complicated for young
children to learn, power-loom weaving developed as a job performed by adolescent girls and young women. It soon became evident that householders could not meet the increasing demand for
these semiskilled operatives. Shortages of weavers threatened production schedules, as Alexander Hodges testified in the 1830s. Slater decided to employ itinerant weavers. Beginning in the mid-
1830s, young, single women entered the factories in increasing
numbers and found work in the weaving room. Sometimes groups
of people applied for work. Louisa Shaw of Globe Village, Southbridge, for example, asked

> if you would like to hire to or three weavers and if so wee would like
> to come and work for you if you would like to have us come wee will
> come as soon as wee can but here they sometimes require thirty days
> notice and sometimes not any but wee will come betwixed now and
> the last of may wee will come or give our notice as soon as we receive
> a line from you I would like to know how much your weavers make
> and how often you settle please sur writ and let mee know yours with
> respect.[13]

Initially most of these workers found accommodation in private
homes or in local boardinghouses, but before the 1840s boardinghouses were few. The Slater firm, for example, operated a small
lodging house and was determined not to expand it. "We wish you
to try to do without the addition to the boarding House this fall

12. Slater MSS, Union Mills, vol. 155, Time Book, 1840.
13. Ibid., Phoenix Thread Mill, vol. 39, Louisa Shaw to Storrs, April 12, 1848;
Union Mills, vol. 190, Patrick Savage to Storrs, June 28, 1851.

partly because . . . the increase of buildings in the village lead us to hope that such an addition may not be permanently necessary," the company advised its agent in 1836.[14] Accommodating itinerant workers with local families could have advantages beyond saving on construction costs. Placed under family government, the young women would be supervised and integrated into industrial community life. This arrangement altered only slightly the employment scheme of having families recruit and accommodate "extra help" on their own; the company now assumed responsibility for hiring some hands and for placing them with local families.

Forces beyond the control of both householders and management further altered the composition of the labor force. Once the backbone of the Slater labor system, children began to be phased out of industrial employment. In the 1820s the Massachusetts legislature, alarmed by reformers' claims that the continued employment of young children would lead inevitably to an ignorant, pauperized, and permanent proletariat, ordered an investigation into child labor. Local officials were asked to gather information on the number, hours of labor, and educational opportunities of persons under the age of sixteen who were employed in incorporated manufacturing companies. Because unincorporated firms were not included in the survey, few replies were received, and no action was taken.[15] A decade passed before the problem received further attention. In response to increased concern over the effects of child labor, a statute passed in 1836 required all children under the age of fifteen who were employed in incorporated manufacturing establishments to attend school at least three months annually. Supporters of the measure in the legislature argued that

if a small part of one generation, however employed, be suffered to become men in physical strength only, without something like a corresponding development of their heads and hearts, their intellects and affections, there is a disease, a canker in the body politic which

14. Ibid., General Box 1, Samuel Slater and Sons to D. W. Jones, September 26, 1836.
15. Susan M. Kingsbury, ed., *Labor Laws and Their Enforcement, with Special Reference to Massachusetts* (New York, 1911), pp. 4–5.

will corrode and spread itself in every direction, to the final destruction of the system.[16]

Because it was organized as a partnership, Samuel Slater and Sons was exempt from this legislation. Nevertheless, Horatio complied with the spirit of the act and allowed his young operatives "the priviledge of three months schooling."[17]

Massachusetts did not enact a minimum age law, but it attempted in 1842 to regulate working hours for children, limiting the workday to ten hours for those under twelve years of age. At first Samuel Slater and Sons tried to evade the law, but in 1848 it was caught for violating the hours act and was taken to court.[18] Soon thereafter Horatio Slater began to replace boys and girls with adolescent and adult laborers. Between 1840 and 1860 the percentage of children twelve and under employed by Slater's textile factories declined from 20 percent to 7 percent.[19] (See Table 12)

Problems associated with the collection of census material, on the one hand, and the ease with which parents and factory masters could deceive census takers about the ages of children and their employment, on the other hand, could cast doubt on the accuracy of these employment figures. Nevertheless, several factors suggest that they probably were correct. By the 1850s labor was plentiful in New England, and Samuel Slater and Sons could hire almost anyone it chose and at a price it was willing to pay. Horatio Slater no longer had to rely primarily on child workers as a source of labor. Furthermore, by reducing child labor in the factories, Slater could achieve one of his goals: he could diminish the power exercised by householders on the factory floor and could assume re-

---

16. Massachusetts, House Document, no. 49 (1836), quoted in Kingsbury, *Labor Laws and Their Enforcement*, p. 19; see also Luther, *Address to the Working-Men of New England*, pp. 10–20.

17. Slater MSS, Slater and Kimball, vol. 3, Agreements with Help, 1827–1839.

18. Ibid., Samuel Slater and Sons, vol. 188, Barton and Bacon to Samuel Slater and Sons, May 13, 1848.

19. Ibid., Union Mills, vol. 155, 1840; Ware, *Early New England Cotton Manufacture*, p. 210; Manuscript Schedules, Seventh Census of the United States, 1850; Eighth Census of the United States, 1860. In Blackstone the percentage of young children twelve and under reported in 1860 was lower, 0.0045. See Manuscript Schedules, Eighth Census of the United States, Blackstone, 1860.

Table 12. Number of workers in 13 occupations, by age group, Webster, 1860

Occupation	12 or less	13– 15	16– 18	19– 24	25– 30	31– 36	37– 42	43– 48	49– 57	58+
					Age group					
Farmer	–	–	–	3	3	6	7	11	16	25
Operative	39	66	103	126	89	41	24	21	17	6
Laborer	2	–	5	17	15	7	18	18	23	16
Shoemaker	1	3	14	44	29	19	14	7	6	6
Carpenter	1	–	–	3	2	3	5	6	2	8
Mechanic	–	–	1	6	3	4	5	2	3	4
Brick mason	–	–	1	6	3	4	1	2	2	1
Merchant	–	–	1	10	11	12	12	6	6	4
Professional man	–	–	–	2	4	3	5	–	–	1
Manufacturer	–	–	1	1	1	2	3	1	1	–
Clerical/managerial	–	–	1	4	2	4	3	3	1	1
Teacher	–	–	3	5	4	1	1	–	–	–
Hotelier	–	–	–	1	1	1	–	1	2	2

Source: Manuscript schedules, Eighth Census of the United States, 1860, Webster, Mass.

sponsibility for the discipline of hands and for the allocation of jobs. State legislation, then, provided Slater with an excuse to re-cruit additional hands, both male and female, and to dilute the influence of householders in the mills.

By the 1850s householders no longer provided all the hands employed by the firm. Very young children were being eased out of factory employment. Itinerant workers migrated to Webster, entered the factories, and worked alongside family labor. In 1851 John Dwyer of Manville wrote: "I am caring [for] 2 boys and 2 girls—1 drawing fender 1 weaver. I want to know what weavers can make for there is 3 more that wants to go there is 1 piecer on mules and one card stripper or a picker tender."[20] In another case, Jane Sillbridge of Willington requested work for herself and for several of her friends. John Carter of Southbridge wanted jobs for himself and for a couple of weavers from the local village. Sometimes laborers from one area moved as a group to Slater's mills. Patrick Savage of Rhodeville had ten or eleven workers in addition

20. Slater MSS, Samuel Slater and Sons, vol. 198, John Dwyer to Samuel Slater and Sons, March 20, 1851.

to himself signed up to transfer to Webster as soon as their employer released them.[21]

Changes under way within the factory would have important consequences for the family system of labor. It was difficult for the Slater family to achieve a higher level of economic rationality while householders continued to influence factory operations. Disciplinary policies, placement procedures, and conditions of labor, customarily part of the domain of householders, were in conflict with the rational organization and operation of the factory floor. Authority could not remain divided between householder and factory master.[22] By the 1830s management appeared to be ready to sacrifice the moral discipline associated with the family and the church in order to obtain more extensive control over the individual worker. Privileges once accorded the householder in the factory came under scrutiny and began to be dismantled as economic factors became the primary influence in the actions of management.

Yet the pace of change was slow; it took almost twenty years for the firm to assume full control over the factory floor. Again, when changes were introduced, disagreements could and did arise, but open conflict usually was avoided. When labor and management disagreed over a proposal, management sometimes postponed the introduction of a new regulation, especially if householders appeared united in their opposition to it. With the passage of time and further changes in conditions, the regulation could be reintroduced. Opposition to some programs was either nonexistent or minimal, and they could be introduced immediately. If a handful of householders opposed a new program they could be dismissed.

Among the first issues addressed by the Slater family was the work schedule. The Slater family introduced Sunday and overtime

21. Ibid., Union Mills, vol. 190, Patrick Savage to Storrs, June 28, 1851; vol. 191, Jane Sillbridge to Storrs, March 1, 1855; Samuel Slater and Sons, vol. 198, John Carter to Samuel Slater and Sons, April 30, 1851.

22. Max Weber, *Economy and Society: An Outline of Interpretive Sociology*, 2 vols., ed. Guenther Roth and Claus Wittich (New York, 1968), 1:137–138.

work, and members of the same family began working different shifts. Some children worked extra hours on Tuesdays and Wednesdays, others on Thursdays and Saturdays; householders and older boys worked on Sundays.[23] Mothers could not be certain when all members of the family or unit would be together. Sunday lost significance as the traditional day for family as well as for religious communion. Further inroads against tradition followed. Morning and afternoon breaks, which had long been periods for workers to meet and chat with kin, to exchange gossip with fellow workers, even to slip out of the mill and dash home or run errands, were abolished. Agents complained that workers took advantage of the breaks, that they stretched the allotted fifteen minutes into forty-five or sixty minutes.[24] By eliminating rest periods, Slater forced more work from his laborers.

Parental supervisory prerogatives also came under attack. Samuel Slater and Sons assumed the power once vested in mule spinners to hire, pay, discipline, and dismiss piecers. A mule spinner no longer had the right to hire and supervise his sons or to teach them his trade. Parents were also forbidden to enter the mill and supervise their children while they were in the overseer's charge. Householders who objected to this regulation were fired. Peter Mayo's entire family, for example, was discharged because he attempted "to control his family whilst under charge of overseer and disorderly conduct generally."[25] Economic incentives and penalties began to replace traditional forms of control within the factory. To encourage acceptable standards of work and behavior, the stick-and-carrot approach was introduced. Black marks were recorded against weavers for shoddy work and fines for tardiness, absence without leave, or disorderly conduct were deducted from their wages, while good work was rewarded by extra allowances.[26]

A further assault against long-standing practices occurred when

23. Slater MSS, Slater and Kimball, vol. 3, Agreements with Help, 1827–1838; Union Mills, vol. 95, 1866; vol. 96, 1861; vol. 155, 1840. Thomas Leavitt, ed., *Hollingworth Letters: Technical Change in the Textile Industry, 1826–1837* (Cambridge, Mass., 1969), p. 22.

24. Joseph Merrill, *History of Amesbury*, p. 366.

25. Slater MSS, Slater and Kimball, vol. 3, Agreements with Help, June 13, 1827; Union Mills, vol. 30, Petty Ledger, 1840–1843.

26. Ibid., Union Mills, vol. 30, Petty Ledger, 1840–1843; Steam Cotton Manufacturing Company, vol. 10, Weaving Accounts, 1829; Slater and Kimball, vol. 3,

manufacturers abandoned the family wage system and began to pay wages directly to individual workers. Initiated in the mid-1830s, this method of payment was first introduced in the weaving department, but when parents complained, the former system was restored. In the early 1840s the firm tried once again, and this time it succeeded. By 1845 each worker received his or her wages. On settlement day in 1845 Daniel Wade, a watchman, and his two adolescent daughters, Laura and Elmira, both weavers, received separate pay slips. Children now could dispose of their own income. With the introduction of the new pay system, contracts were eliminated.[27]

When the householder collected all wages, he controlled the available income and distributed it according to his priorities. With this new arrangement, however, economic power shifted in part to children and adolescent wage workers, and the householder's domination of the family was threatened. Parents had to negotiate with each child over the disposal of his or her wages. With economic independence, with jobs available to individuals in Webster and elsewhere, children, charges, and boarders could move out of the family home and take up residence in local boardinghouses or leave the community altogether.[28] By this time the company operated two large factory boardinghouses, one that accommodated fifty-six men and women and another that housed twenty-eight people. Local residents also took in lodgers.[29]

Agreement with Weavers, 1836 and 1837; see also Obadiah Greenhill, 1827, and Abby Pierce, 1833.

27. Ibid., Slater and Kimball, vol. 3, Agreements with Help, 1827–1840; Union Mills, vol. 30; vol. 175, December 1845; vol. 170, 1836.

28. The family wage system and its implications for householders are discussed in Anderson, *Family Structure in Nineteenth-Century Lancashire*, pp. 130–132. It is difficult to say precisely when Slater began to pay hands directly. There is evidence that some family workers left home and moved to other industrial communities where they would be able to collect their own wages. See Slater MSS, Slater and Kimball, vol. 3, Agreements with Help, 1827–1840; Webster, Mass., Vital Statistics, Marriages, 1851; Manuscript Schedules, Seventh Census of the United States, 1850; Slater MSS, Union Mills, vol. 30, Petty Ledger, 1840–1843, especially the family of John McCausland.

29. Slater MSS, Union Mills, vol. 190, Young to Storrs, January 27, 1846; vol. 121, Cole to Storrs, January 19, 1858; vol. 190, D. Sewell to Storrs, March 18, 1851. Some people took their meals at the boardinghouses; see vol. 182, September 20, 1851; August 1, 1853; Manuscript Schedules, Seventh Census of the United States, 1850.

Families began to drift apart. The family of Thomas Jepson did not make it through the era intact. With his wife and four children, Jepson arrived in Webster in 1848. Probably he took a farm on shares, and he placed two of his daughters, Amanda and Amy, eighteen and thirteen years of age, in the factories. The two girls were able to support themselves. After living at home for approximately a year, they moved into a factory boardinghouse at North Village. In 1854 Amanda Jepson married Amos White of Westfield, Connecticut, and her sister married George Shumway of Woodstock, a harness maker.[30]

Families that had lived in the community for decades also were affected. John McCausland, a founder of the Methodist church in Webster, saw his children pursue their own economic self-interest. In the 1820s McCausland, his wife, and their six children arrived in the Oxford South Gore area. He secured a job as a watchman for Slater and Kimball in 1827 and placed three of his children, Alexander, Jane, and James, in the factory. Alexander learned to operate spinning mules and left Webster shortly thereafter. Jane, who started working in the drawing and trimming department, transferred to the weaving room. By the early 1840s she too left Webster and moved from one factory town to another until 1851, when she married James Apperson, a mason she met while living in Southbridge. James McCausland remained with the firm until the 1850s. In 1844 he became an overseer in the carding room at the Phoenix Thread Mill and earned 6s. a day. Soon thereafter he left his father's home and moved into a company boardinghouse. All of the children who went to work with their father in 1827 left home before they married; indeed, two of them moved away from the community.[31]

Within the factory the family unit was broken down; individual

30. Manuscript Schedules, Seventh Census of the United States, 1850; Webster, Mass., Vital Statistics, Marriages, 1854.

31. Between 1831 and 1836 three other children, John Jr., Elizabeth, and Mary, entered the factories. Mary, age eight, died the same year she started work; John Jr. remained at home until he married in July 1850, and Elizabeth remained at home with her parents until John McCausland died in the mid-1850s. See Slater MSS, Slater and Kimball, vol. 3, Agreements with Help, 1827–1840; Webster, Mass., Vital Statistics, Marriages, 1851, 1850; Webster, Mass., Vital Statistics, Deaths, 1836; Manuscript Schedules, Seventh Census of the United States, 1850.

laborers rather than the work unit, became the focus of the factory agent, who dealt directly with each member and no longer communicated through the householder. With this development the cash nexus began to replace traditional forms of control, and authority passed increasingly from the householder to the factory agent. Economic considerations formed the basis of the new relationship between Slater and the operatives.

While Slater checked the power of the householder, he also tried to introduce new policies that would increase the productivity of his labor force. The workday was stretched another fifteen, twenty, or thirty minutes, depending on the whim or the needs of the factory agent. It should be remembered that few workmen had clocks, and that a factory bell summoned hands to work, signaled breaks for lunch and dinner, and tolled again at quitting time. But factory time invariably fell behind true time, and agents exacted extra work from hands. This was the case not only in the Slater mills but throughout southern New England. At the Hope factory in Rhode Island, operatives started work approximately twenty-five minutes after daybreak and did not leave for home until the factory bell signaled the end of the day at 8:00 P.M. But as the *Free Inquirer* reported, 8:00 by factory time "is from twenty to twenty-five minutes behind the true time."[32] Many manufacturers defended this practice, arguing that "the workmen and children being thus employed, have no time to spend in idleness or vicious amusements."[33] Forced to maintain the production schedules set by the manufacturer and to cut costs where possible, supervisors lengthened the workday to obtain additional labor from hands.

To increase production further, the stretch-out and the speed-up were intensified. Slater crowded more and more machines into already cramped spinning, carding, and weaving rooms and assigned additional machines to each worker. In the weaving department, for example, the number of looms attended by each weaver

32. *Free Enquirer,* June 14, 1832, quoted in Commons, ed., *Documentary History of American Industrial Society,* 5:196.
33. Smith Wilkinson to George White, n.d., quoted in White, *Memoir of Samuel Slater,* p. 126; see also Luther, *Address to the Working-Men of New England,* p. 20; *Paterson Courier,* August 11, 1835, quoted in Commons, ed., *Documentary History of American Industrial Society,* 5:64.

was increased steadily from two to three to four and then to six.[34] Not only did the hands operate more machines, but the machines were run at higher speeds. Initially the speedup was management's response to the pressure of weavers who worked on a piece-rate basis (approximately 20 cents a cut for weaving 4-by-4 sheeting and shirting), who could increase their earnings only by producing more cloth. In 1837 Alexander Hodges complained to the head office: "The weavers are being uneasy about the speed being slow and some of the new ones will leave. I think we had better put on a little in order to keep the best of them nothing short of this will answer as the mills in this vicinity have advanced the prices."[35] Soon, however, the speedup became a method to increase production. Writing to Union Mills in 1855, Fletcher confided, "As soon as the supply of good weavers can be obtained, . . . the increased speed of looms will show itself by increased quantity."[36] Piece rates remained constant at the earlier level. The speedup and stretch-out were introduced into almost every room in the factory, without an appreciable increase in pay.[37] (See Table 13.)

Supply and demand factors, new technology, education laws, and factory acts worked together by the 1850s to transform the labor force employed at Samuel Slater and Sons. The route pursued by the firm led toward an autonomous worker, one who was cut off from his or her family and looked to the factory system for opportunity, support, and survival. The labor force had been stream-

---

34. Slater MSS, Slater and Kimball, vol. 3, Agreements with Help, 1827–1840; Union Mills, vol. 190, Ruth Graves to Storrs, March 12, 1846; vol. 191, Jane Sillbridge to Storrs, March 1, 1855; vol. 186, A. Hodges to Samuel Slater and Sons, November 4, 1837; vol. 189, Fletcher to Union Mills, January 8 and August 22, 1855; and vol. 188, Fletcher to Union Mills, November 5, 1852; vol. 114, Storrs to C. Hale, August 6, 1858; vol. 117, Storrs to Horatio Nelson Slater, May 5, 1845; vol. 187, Fletcher to Union Mills, January 9, 1852; see also "Two Hundred Years of Progress," pp. 8–9.

35. Slater MSS, Union Mills, vol. 186, A. Hodges to Samuel Slater and Sons, November 4, 1837.

36. Ibid., Union Mills, vol. 189, Fletcher to Union Mills, August 22, 1855; Horatio N. Slater, vol. 33, S. Levalley to Horatio N. Slater, February 20, 1837.

37. Ibid., Union Mills, vol. 188, Fletcher to Union Mills, November 5, 1852. Although Samuel Slater and Sons did not increase wages, it also held constant certain charges. Between 1840 and 1861 room-and-board rates for single women rose from $1.33 to $1.38 per week. See Slater MSS, Union Mills, vol. 30, April 1840–March 1841; vol. 182, Orrel Bigelow Boarding House, October 22 to November 18, 1854; vol. 193, North Village Boarding House, December 7, 1861.

Table 13. Minimum and maximum wage rates paid by Union Mills, 1840–1857

Occupation or department	1840		1847		1852		1857	
	Min.	Max.	Min.	Max.	Min.	Max.	Min.	Max.
Spinning per week	6s.	15s.6d.	6s.	18s.	6s.	24s.	8s.	24s.
Carding per week	6s.	16s.6d.	9s.	21s.	6s.	20s.	9s.	24s.
per day	–	–	–	–	4s.	9s.6d.	–	–
Weaving 4/4 cloth, per cut	$0.20	$0.20	$0.18	$0.18	$0.21	$0.23	$0.19½	$0.20½
Mule spinner per 100 skeins	$0.09	$0.10¾	$0.06	$0.06	$0.05	$0.05½	$0.04¾	$0.05
Piecer per week	6s.	14s.	6s.	14s.	6s.	16s.6d.	6s.	24s.
Dressing per cut	$0.04	$0.04	–	–	$0.02½	$0.03	–	–
per day	–	–	–	–	–	–	7s.7d.	9s.
Machine shop per day	4s.	8s.	3s.6d.	8s.*	7s.6d.	7s.6d.	8s.	10s.6d.

Note: Those who were learning to operate a machine received no wages during the training period. In the carding room it took approximately three days to learn to operate the equipment.
*Or by the week.
Source: Slater MSS, Union Mills, vol. 30, 1840–1843, Petty Ledger; vols. 144, 147, 149, Time Books, 1840, 1843–1857.

lined; the family unit, tied to a firm social base, was replaced by an individual tied to the wage economy.

FLIGHT

Modifications introduced by Samuel Slater and Sons in the ante-bellum era had an unsettling effect on native-born Americans and their families. Many of them could not accept their new economic and social positions in the factories and in the community. Yet the options for redress were few: they could join together against their employer and resist the new changes, or they could leave. Given their background, their values, and their long-standing relationship with the manufacturer, many of them would have found it difficult to confront the firm. No matter how harsh conditions had become, no matter what misfortunes befell them, no matter how angry they might have been, many of these native-born householders could not openly defy Samuel Slater and Sons. While weakened, the tie that bound labor to management was still in evidence. According to Barrington Moore, Jr., in order for the laborer to sever this tie, there would have had "to be experiences along the way enabling him to overcome his affectionate dependence on paternalistic authority, or creating in him a very different attitude toward figures of authority."[38] Rather than weaken authority or challenge patriarchy, these workers sought to preserve traditional values. To question authority in the workplace would have been to admit that it might no longer be legitimate; such questioning might have opened the door to challenges in the home or in the church. Flight, not overt conflict, represented a more viable solution to the workers' changed circumstances. While economic conditions now determined management decisions, customary beliefs and traditional values still influenced the judgment, choices, and actions of native-born residents. Changes in factory policy and alterations in the composition of the work force from 1830 to 1860 went hand in hand with high rates of labor turnover and increased out-migration, suggesting that workers did indeed choose flight rather than resistance.[39]

38. Moore, *Injustice*, p. 125.

39. The information presented here and in the discussion on mortality is based on a variety of sources, including federal population census manuscript schedules for 1840, 1850, and 1860; parish records for the Methodist and Congregational churches; company payroll ledgers from 1827 to 1860; and vital statistics for Webster from 1832 to 1860. Although the Commonwealth of Massachusetts required all local communities to record information on births, deaths, and marriages, the early records are somewhat suspect. In the Jacksonian era responsibility for reporting births, for example, rested with parents, and they often failed to notify local officials, especially if a child was stillborn or died within a few days of birth. Similarly, deaths were underreported, and the cause of death was often not stated. Records began to improve in the 1840s, when age and cause of death of individuals were reported with more regularity. After 1844 records improved, and by 1860 approximately 92 percent of all deaths were recorded. The material collected from all of these sources was arranged according to the family reconstitution method; altogether almost 7,000 individuals were counted in the survey. Although every attempt was made to gather information, the arguments based on these data are more suggestive than conclusive. Only people who left a written record of their stay in Webster were recorded, and of course many came and went without a trace. See Maris A. Vinovskis, "Mortality Rates and Trends in Massachusetts before 1860," *Journal of Economic History* 32 (March 1972):184–192; Robert Gutman, "Birth Statistics of Massachusetts during the Nineteenth Century," *Population Studies* 10 (July 1956):74–78. For an understanding of the method employed to construct the survey, see Ed-

Because Webster was not incorporated until 1832, and because during the 1830s the vital statistics collected for the community were incomplete or unreliable, a detailed demographic outline of Webster cannot be drawn for that decade. Yet an indication of the degree of commitment to the Slater firm and to the village can be ascertained by examining employment records for the Slater and Kimball factory, an early Slater mill. In 1829 approximately eighty-nine people worked for this factory; this figure includes seven householders who held skilled or management positions and twenty-one family units. The twenty-eight householders who either worked for Slater directly or placed their children or charges in the mills formed the basis for this survey. Of the twenty-eight householders listed on the worker's ledger in 1829, sixteen, or 57 percent, remained in Webster at least one decade; two died during the 1830s and ten left the community. Of the sixteen householders who remained in the village, fourteen and/or their families continued to work for Samuel Slater and Sons throughout the 1830s and could be found on the ledger books in 1840. Two householders left the mills altogether and turned to agriculture for their livelihood. Of those who left the mill and the village, four families had worked for Slater and Kimball for less than two years, five for three to four years, and one for five years. Although the number of households surveyed was small, this sample suggests that factory families in the 1830s not only remained in the community for a considerable length of time but continued in factory employment for many years.[40]

ward Wrigley, ed., *An Introduction to English Historical Demography* (New York, 1966), and his "Family Limitation in Pre-Industrial England," *Economic History Review* 19, 2d ser. (1966):82–109; and David Levine, *Family Formation in an Age of Nascent Capitalism* (New York, 1977).

For the material included in the family reconstitution survey for Webster, see Webster, Mass., Vital Statistics, Births, Deaths, Marriages, 1832–1860; United Church of Christ MSS, "Methodist Episcopal Church at the Four Corners," and Records of the Quarterly Meeting Conference of Webster; *Historical Sketch and Directory of the Methodist Episcopal Church, Webster, Massachusetts* (1904). See also United Church of Christ MSS, Congregational Church Parish Records, 1838–1911; Daniels, *History of the Town of Oxford*, and Manuscript Schedules, Sixth Census of the United States, 1840; Seventh Census of the United States, 1850; Eighth Census of the United States, 1860.

40. Slater MSS, Slater and Kimball, vol. 3, Agreements with Help, 1827–1840; Webster, Mass., Vital Statistics, Marriages, 1832–1840; United Church of Christ MSS, Congregational Church Parish Records, 1838–1911; Manuscript Schedules, Sixth Census of the United States, 1840; Slater MSS, Union Mills, vol. 30, Petty Ledger, 1840–1843.

During the 1830s the population of Webster registered a modest gain; it rose from 1,170 in 1832 to approximately 1,400 by 1840, the date of the first federal census in Webster Township.[41]

This pattern of persistence held for the next decade. Because of the availability of records, the survey for the 1840s is larger and more complete. In 1840, 236 male householders lived in Webster. Of these men, 139, or approximately 58.5 percent, remained in the community through the decade. The rate of persistence from 1840 to 1850 was comparable to the rate calculated for the Slater and Kimball sample a decade earlier.

Although undoubtedly deaths sometimes went unlisted, the town clerk reported that 23 male householders, or 9.7 percent of those resident in 1840, died in Webster that decade. (See Table 14.) Many died of "old age," but a significant number of people contracted diseases endemic in Webster, including tuberculosis and typhus. Transmitted from person to person by the body louse, typhus fever reached epidemic proportions during the 1840s. It took many lives in 1843 and appeared again in 1844 and 1846. Women and adolescents were the primary victims.[42]

Table 14. Causes of death of male householders, Webster, 1840–1850 and 1850–1860

Cause of death	1840–1850	1850–1860
Apoplexy	2	1
Congestion	0	1
Consumption or tuberculosis	6	6
Delirium tremors	0	1
Diabetes	0	1
Dropsy	1	0
Dysentery	1	0
Erysipelas	2	0
Lung fever	0	1
Numb palsy	1	0
Old age	7	13
Sunstroke	0	1
Typhus	2	0
Unknown	1	4

Source: Webster, Mass., Vital Statistics, Deaths, 1840–1860.

41. "Two Hundred Years of Progress," p. 4; see also Manuscript Schedules, Sixth Census of the United States, 1840.

42. See René Dubos and Jean Dubos, *The White Plague: Tuberculosis, Man, and Society* (London, 1952), pp. 44–66, 117, 169, 219; Richard H. Shryock, *Medicine and*

The amount of out-migration in the 1840s was small. Approximately seventy-four householders were known to have left Webster between 1840 and 1850. Representative of those who left were Charles and Roxalana Goulding. Around 1838 the Gouldings and their six children had moved to Webster from West Boylston. Four of the children, including young Charles, George, Henry, and William, found work in the carding and spinning departments at the Phoenix Thread Mill. Goulding secured a job as a teamster and part-time blacksmith. Throughout the depression years the family remained in Webster, but they left for Worcester following the death of the youngest son, Emanuel, in 1845.[43] Families such as the Gouldings appeared to move within a short radius. While it is difficult to determine with accuracy the destination of these factory families, an impression can be gained from the Congregational parish records for the period. The minister of the church recorded both the location of a family's last parish and its destination when it left Webster. It appears that families moved from one industrial community to another. Dudley, Oxford, Thompson, Sturbridge, Whitinsville, Hadley Falls, Southbridge, and Worcester were the recorded destinations of many of those who left Webster.[44]

Out-migration accounted for 31.4 percent of Webster's families during that decade, and those who remained in the community did not necessarily remain employed by Samuel Slater and Sons. Some workers quit the factories and entered the boot and shoe industry, agriculture, or commerce. During the 1840s Webster attracted new industries. Henry Bugbee opened a small office in the basement of a local store in 1843 and set up as a shoe manufacturer. When he started his business, shoemaking was still in the household stage of production; he sent his leather to Natick to be cut and then put it

---

Society in America, 1660–1860 (New York, 1960), p. 95; John Duffy, *Epidemics in Colonial America* (Baton Rouge, 1953), pp. 215–218; George Rosen, "Disease, Debility, and Death," in *The Victorian City*, ed. H. J. Dyos and Michael Wolff, 2 vols. (London, 1973), 2:642–643; and Carl Bode, *The Anatomy of American Popular Culture, 1840–1861* (Berkeley, 1959), p. 175.

43. United Church of Christ MSS, Congregational Church Parish Records, June 24, 1838; Slater MSS, Phoenix Thread Mill, vol. 27, March 1842; vol. 28, June 1839, January 1840, February and September 1844; Manuscript Schedules, Sixth Census of the United States, 1840; Webster Mass., Vital Statistics, Deaths, June 22, 1845; and Slater MSS, Union Mills, vol. 30, Petty Ledger, 1840–1843.

44. United Church of Christ MSS, Congregational Church Parish Records, 1838–1911.

out to local families to bound and bottomed. Women usually performed the former tasks and men the latter. Bugbee's business prospered, and others joined the industry. By 1850 sixty-nine people earned their living as shoemakers.[45] Some of those who turned to shoemaking were like Alonzo Larned, who had worked in the mills for years. The Larned family had arrived in Webster in the mid-1830s. Alonzo found a job as a machinist at Union Mills, where he earned 5s. 6d. a day, or approximately $22 a month. In 1840, along with several other men, Alonzo Larned applied for an overseer position in the carding room and was given a one-month trial. He was not selected for the post, however, and he returned to the machine shop. The following year another opening occurred, this time in the carding room at the Phoenix Thread Mill, for a second hand. Larned applied and was hired; later, in the spring of 1842, he was promoted to overseer and received a slight increase in pay, to 6s. a day. He remained there for several years, but in April 1844 he left Slater to become a shoemaker.[46] Unfortunately, neither the company records nor personal letters indicate why Larned chose to leave the firm, but the stories of other operatives who turned to this domestic industry for support are available.

In the early years of industrialization, for many young women a change of marital status called for a change of job. Most women ceased factory work when they married. Even if they wanted or needed to work, societal pressure forced them out of the factories and, as often as not, into another line of work. The Foster sisters, Lucy and Caroline, were representative of the many women who spent their youth in the mills and found other remunerative employment once they married. The daughters of Ebenezer Foster, a local farmer, they entered the Slater factories in the 1830s. By 1840 both sisters worked as weavers at Union Mills. In 1844, at the age of twenty-four, Lucy Foster married a Southbridge, Massachusetts,

45. "Two Hundred Years of Progress," p. 8; Slater MSS, Union Mills, vol. 155, Time Book, 1840; Slater and Kimball, vol. 3, Agreements with Help, 1827–1836; Daniels, *History of the Town of Oxford*, p. 512; Webster, Mass., Vital Statistics, Marriages, 1844 and 1845; Manuscript Schedules, Seventh Census of the United States, 1850; Eighth Census of the United States, 1860.

46. Slater MSS, Union Mills, vol. 144, July 1840; Phoenix Thread Mill, vol. 27, March 1842–January 1843; vol. 28, January 1844–April 1844; Slater and Kimball, vol. 3, Agreements with Help, John McCausland, 1827; "Two Hundred Years of Progress," p. 8; Manuscript Schedules, Seventh Census of the United States, 1850.

schoolteacher, Francis Davis, and within a year Caroline Foster, twenty-two years of age, married Barlow Hoyle, a young operative employed by the Slater firm. Shortly after their marriages the women left the factory, and subsequently both of their husbands became shoemakers, an occupation that could be performed in the home and that could make use of the labor of women. Apparently the family unit could not survive on a single salary; the women had to work. For both families shoemaking represented an effective compromise between social pressures, which militated against the employment of married women outside the home, and economic necessity.[47]

During the 1840s opportunities were available in agriculture. Many factory families longed to work a piece of land of their own. Few could afford to purchase land in the area, however; most rented a farm on shares from the Slater firm and raised corn, rye, and potatoes. While tenants paid an annual fee for the land, the farmer and the firm shared expenses for seed, fertilizer, and taxes. The firm chose its tenants with care, and many of them and/or their families had worked for the firm in some capacity. Lewis Shumway, for example, who rented a farm for $458, sent four of his children to work in the mills. Subsequently all of them left the factories: two remained in Webster and worked with their father on the farm, and the other two left the village.[48]

While Shumway had not worked in the factories himself, others who turned to agriculture had industrial experience. Rufus Freeman, Reed Smith, and William Tanner, all overseers and second hands at Union Mills in 1840, left the factories that decade to become farmers. Born in Massachusetts, Rufus Freeman and his wife, Clarissa, arrived in Webster during the late 1830s. At the age of twenty-five he began supervising the weaving room, where he remained until 1846. While the amount of money Freeman was able

47. Daniels, *History of the Town of Oxford*, p. 512; Manuscript Schedules, Eighth Census of the United States, 1860. While both women remained in Webster throughout the 1840s and 1850s, Caroline and Barlow Hoyle eventually left town and settled in Detroit. Lucy and Francis Davis remained in the community and raised their two sons there.

48. Slater MSS, Union Mills, vol. 30, Petty Ledger, 1840–1843; Union Mills, vol. 101, Leasing Agreement, Augustus Emerson, April 27, 1837; vol. 87, Old Rent Book, Howland Farm, April 1, 1849, B. F. George and Lewis Shumway.

to save fluctuated widely from one year to the next, the accounting period from March 1841 to April 1842 was especially good for him. In March 1841 he received a cash settlement of $140, and in the spring of the following year he received another settlement of $313. By 1846 Freeman had saved enough money to take a farm. At that time he was married and had a six-year-old daughter and a baby son. By 1850 he claimed assets worth $1,000.[49] William Tanner, who in 1839 married Hannah Slater, a mill worker, served as an overseer at Union Mills in the 1840s. Early that decade he took a farm on shares. Neither Tanner nor Freeman, however, remained in agriculture. After a trial period, they returned to the factory in their former capacities. Only Reed Smith, whose father owned a farm in the area, succeeded in agriculture.[50]

For Webster the 1840s was a decade of growth. The population rose from 1,400 to 2,361, an increase of 61 percent. Although immigrants had begun to enter the community, the bulk of the population was still native-born. Such people dominated the factories and filled most of the unskilled and semiskilled jobs in Webster. While they may have moved with more frequency from one job to another during this decade than they had in the previous one, they seem to have considered Webster a desirable place to live and the factories an acceptable place to work.[51]

Beginning in the 1850s persistence rates dropped. Of 303 native-born families who lived in Webster in 1850, only 121, or 40 percent, could be found there a decade later.

Approximately 29 householders, or 9.57 percent, died in the 1850s. (See Table 13.) This rate approximated that of the 1840s, although it should be interpreted with caution. While many men died of "old age" in Webster, and indeed the proportion of the native-born population over the age of sixty rose, disease claimed lives. Tuberculosis, known as the "white plague," which often sub-

49. Slater MSS, Union Mills, vol. 30, Petty Ledger, March 1841–April 1842; Webster, Mass., Vital Statistics, Births, 1840, 1846; Manuscript Schedules, Seventh Census of the United States, 1850; Sixth Census of the United States, 1840.

50. Slater MSS, Union Mills, vol. 155, Time Book, 1840; Webster, Mass., Vital Statistics, Marriages, 1839; Manuscript Schedules, Sixth Census of the United States, 1840; Seventh Census of the United States, 1850.

51. Slater MSS, Samuel Slater and Sons, vol. 196, Thomas Midgley to Samuel Slater and Sons, July 21, 1845.

jected its victims to long periods of illness and slowly drained their vitality, was widespread. Little notice was taken of the disease and scarcely any effort was made to control it. Spread by close, regular contact between people over a period of weeks or months, tuberculosis flourished in the dwellings and the workshops and factories in Webster, taking the lives of six householders in the 1850s, as it had done in the 1840s.[52]

Out-migration became significant in this decade. Approximately 50 percent of those listed on the 1850 census rolls left the community. In fact, the native-born population of Webster barely maintained itself. Between 1850 and 1860 the native-born population increased by only 43 people. The net natural increase that decade was 249 people.[53] The small increase in the number of native-born residents masks the true dimension of population change and turnover. Peter Knights noted that population turnover "was much greater than censuses would lead us to believe, because each population stream had a counter-stream, and large total movements produced small net population changes."[54] Population movements are difficult to estimate, but the data base collected for Webster, which includes everyone who left a written record of residence in the town between 1840 and 1860, reveals that approximately 3,066 native-born Americans resided in the community at some time between 1850 and 1860. If the native-born population totaled 1,873 in 1860, then approximately 1,193 moved in and out of the community that decade.[55]

Between 1830 and 1860 native-born workers left the factories. While the rate of out-migration was low in the earlier decades, it

52. See Dubos, *White Plague*, pp. 44–66, 117, 169, 219.
53. Manuscript Schedules, Seventh Census of the United States, 1850; Eighth Census of the United States, 1860; Webster, Vital Statistics, Deaths, 1850–1860; Births, 1850–1860. Net natural increase was calculated by subtracting the number of deaths of native-born people (531) from the number of births in Webster (780).
54. Peter R. Knights, *The Plain People of Boston, 1830–1860: A Study in City Growth* (New York, 1971), p. 53.
55. This estimate for minimum population turnover is only slightly lower (fifty people) than that calculated by the formula suggested by Robert Doherty, *Society and Power: Five New England Towns, 1800–1860* (Amherst, 1977), p. 105 n. 4.2; in Warren, a Rhode Island mill town situated on Narragansett Bay, rapid out-migration occurred during the same period. The population declined from 3,103 to 2,636. See Frank Mott, "Portrait of an American Mill Town: Demographic Response in Mid-Nineteenth-Century Warren, Rhode Island," *Population Studies* 26 (1972):151.

accelerated in the 1850s until by 1860 native-born workers formed only 28 percent of the entire textile labor force and only 65 percent of the population of the community.[56] The exodus of native-born workers from mill and community coincided with major changes introduced by the Slater brothers in the factory to cut costs and to limit the influence of householders. Unable to accept the changes and unwilling to strike or to take other overt action against Samuel Slater and Sons, workers left.

NEW INDUSTRIAL WORKERS

Although the increase in the native-born population was small, the population of the community grew substantially during the 1850s. By 1860 Webster's population totaled 2,898, and approximately 35.4 percent of the residents, or 1,025 people, were foreign-born.[57] Immigrants moved into Webster and took the place of native-born workers in the factories.

The trend toward the employment of immigrant workers had begun in the 1840s, as increasing numbers of foreign-born laborers found work in the Slater factories, and it accelerated in the decades that followed. By 1850 immigrants from Ireland, Germany, Scotland, England, and Canada comprised 38 percent of the textile labor force there. (See Table 15.) Ten years later they dominated the work force, commanding approximately 72 percent of all jobs in the factories. Of the 532 operatives listed on the federal census manuscript schedules for Webster in 1860, only 147 workers were native-born Americans; 208, or approximately 39 percent, were French-Canadian, and another 144, or approximately 21 percent, were Irish.[58] (See Table 16.) Other industrial communities experienced similar changes in the ethnic composition of the work force. In Blackstone by 1860, immigrants comprised 84 percent of all textile workers, 51 percent of domestic servants, and 62 percent of the unskilled casual and day hands.[59] (See Table 17.)

56. Manuscript Schedules, Eighth Census of the United States, 1860.
57. Ibid.
58. Manuscript Schedules, Seventh Census of the United States, 1850.
59. Manuscript Schedules, Eighth Census of the United States, 1860, Blackstone, Mass.

Table 15. Number of workers in 15 occupational categories, Webster, 1850, by state or country of origin

State or country of origin	Laborer	Farmer	Operative	Domestic servant	Shoemaker	Manufacturer	Merchant	Teacher	Carpenter	Mechanic	Hotelier	Clerical/managerial	Professional	Brick mason	Miscellaneous
None given	1	—	2	1	—	—	—	—	—	—	—	—	—	—	—
Massachusetts	23	66	107	36	49	3	18	3	18	13	1	9	5	9	3
Connecticut	6	13	55	12	17	—	7	—	3	3	—	3	3	7	2
Vermont	1	4	7	1	1	—	1	—	—	2	—	—	—	—	1
New Hampshire	—	2	—	—	—	—	—	—	—	—	—	—	—	—	1
Rhode Island	2	7	16	4	2	—	2	—	3	2	1	4	—	2	—
New York	—	2	—	—	2	—	—	—	1	2	—	1	1	—	—
Canada	47	—	14	—	1	—	—	—	—	2	—	—	—	—	—
Germany	—	—	29	—	—	—	—	—	—	—	—	—	—	—	—
England	4	1	22	2	—	1	2	—	—	—	1	—	—	1	2
Scotland	1	—	5	1	—	—	—	—	—	—	—	—	—	—	—
Ireland	44	—	44	2	2	—	2	—	—	—	—	—	—	1	—
Switzerland	—	—	—	—	—	—	—	—	—	—	—	—	—	—	—
	129	95	301	59	75	4	32	3	25	24	3	17	9	20	9

Source: Manuscript schedules, Seventh Census of the United States, 1850, Webster, Mass.

Table 16. Number of workers in 15 occupational categories, Webster, 1860, by state or country of origin

State or country of origin	Operative	Laborer	Shoemaker	Domestic servant	Carpenter	Mechanic	Clerical/managerial	Brick mason	Merchant	Professional	Teacher	Hotelier	Manufacturer	Farmer	Miscellaneous
None given	3	4	3	—	1	—	1	1	1	—	—	—	—	1	1
Massachusetts	79	23	77	6	10	12	9	4	27	8	11	5	9	47	3
Connecticut	45	9	16	1	6	5	5	9	14	1	2	1	1	15	1
Vermont	9	3	1	—	—	—	—	1	2	—	1	—	—	1	1
New Hampshire	2	—	2	—	—	1	1	—	1	2	—	1	—	—	—
Rhode Island	6	4	6	—	7	5	3	1	1	—	—	1	—	1	—
New Jersey	6	—	4	1	1	1	—	—	2	—	—	1	—	4	—
New York	—	—	—	—	—	—	—	—	—	1	—	—	—	—	—
Canada	208	43	3	—	5	1	—	1	—	—	—	—	—	—	—
Germany	29	2	8	1	—	—	—	2	1	—	—	—	—	—	—
England	25	1	3	—	—	—	—	—	3	2	—	—	—	1	—
Scotland	6	—	—	—	—	—	—	—	—	—	—	—	—	—	—
Ireland	114	31	20	14	—	2	—	1	9	—	—	—	—	1	—
Italy	—	1	—	—	—	—	—	—	—	—	—	—	—	2	—
Switzerland	—	—	—	—	—	—	—	—	1	—	—	—	—	—	—
	532	121	143	23	30	28	19	20	62	15	14	8	10	71	6

Source: Manuscript schedules, Eighth Census of the United States, 1860, Webster, Mass.

Table 17. Numbers of workers in 16 occupational categories, Blackstone, 1860, by state or country of origin

State or country of origin	Farmer	Operative	Laborer	Merchant	Professional	Shoemaker	Carpenter	Manufacturer	Mechanic	Clerical/ managerial	Domestic servant	Teacher	Hotelier	Brick mason	Skilled operative	Miscellaneous
None given	—	5	4	2	—	2	1	—	—	—	1	1	—	—	—	—
Massachusetts	54	91	97	63	16	47	21	5	9	3	27	28	1	9	17	29
Connecticut	—	9	3	4	1	3	1	1	2	1	2	2	—	—	4	4
Vermont	—	5	1	4	1	2	—	—	1	—	3	1	—	—	1	2
New Hampshire	—	8	3	4	—	1	1	—	1	—	2	2	—	1	—	2
Rhode Island	10	27	23	32	4	8	7	5	9	2	8	11	—	1	3	11
New York	—	18	1	3	—	1	2	—	1	—	—	1	1	—	—	1
Pennsylvania	—	—	—	1	—	—	—	—	—	—	—	1	1	—	—	—
New Jersey	—	—	—	1	—	—	—	—	—	—	—	—	—	—	—	1
Maine	—	9	9	5	2	1	1	—	—	—	2	1	—	1	1	1
Canada	—	12	5	—	1	—	1	—	—	—	—	—	—	—	—	—
Germany	—	6	1	2	—	—	—	—	—	—	1	—	—	—	1	—
England	2	129	11	13	—	2	1	—	1	1	1	1	1	2	15	5
Scotland	—	8	3	2	—	—	—	—	—	—	—	—	—	—	1	2
Ireland	2	765	207	41	7	13	5	—	3	—	44	6	—	14	50	18
	68	1,092	367	177	32	80	41	11	27	7	92	55	4	28	93	76

Source: Manuscript schedules, Eighth Census of the United States, 1860, Blackstone, Mass.

In both Blackstone and Webster, the Irish were the first to enter the factories. On arrival in Massachusetts, many Irish immigrants found work in the textile mills at Lowell, Holyoke, Lawrence, Chicopee, and Waltham, where by midcentury they comprised respectively 28, 33, 30, 33, and 28 percent of the total population. In the industrial communities of Ware and Blackstone, where family labor had prevailed, the Irish comprised 22 and 37 percent, respectively, of the total population.[60] But Webster attracted far fewer Irish than these communities. The Irish population increased from 150 in 1847 to 188 at midcentury (see Table 18) and to 375 a dec-

Table 18. Population of Webster, 1850, by state or country of origin

State or country of origin	Number
None given	6
Massachusetts	1,330
Connecticutt	305
Vermont	34
New Hampshire	12
Rhode Island	105
Virginia	1
Ohio	1
New York	29
Pennsylvania	6
New Jersey	1
All American-born	1,830
Canada	189
Germany	61
Switzerland	5
England	77
Poland	1
Scotland	10
Ireland	188
All foreign-born	531
Total population	2,361

Source: Manuscript schedules, Seventh Census of the United States, 1850, Webster, Mass.

60. *Abstract of the Census of the Commonwealth of Massachusetts with Reference to Facts Existing on the 1st Day of June, 1855* (1857); Dublin, *Women at Work*, p. 170; Ware, *Early New England Cotton Manufacture*, p. 230; Massachusetts, Senate, *Report of the Hon. H. K. Oliver, Deputy State Constable Especially Appointed to Enforce the Laws Regulating the Employment of Children in Manufacturing and Mechanical Establishments; with Notice of Operative and Adult Child-Life, and Wages in Europe and in Massachusetts*, S. no. 44, 1868, p. 11; Constance Green, *Holyoke, Massachusetts: A Case History of the Industrial Revolution in America* (New Haven, 1939), p. 30; and Manuscript Schedules, Eighth Census of the United States, 1860, Blackstone, Mass.

ade later. By 1860 Irish immigrants made up slightly less than 13 percent of the total population.[61] (See Table 19.)

To the "famine Irish," Webster offered limited economic opportunities. Many arrived as gang laborers on the Norwich and Worcester Railroad, and some wanted to remain in Webster and make it their home. But these new arrivals, most of them young, unmarried men, found it difficult to secure employment in a community still dominated in 1840 by family labor. Although some continued to work for the railroads and others were hired by local farmers and shopkeepers, most remained in Webster only a few months before moving to another village.[62] With the decline of child labor in the late 1840s, however, opportunities opened up for these

Table 19. Population of Webster, 1860, by state or country of origin

State or country of origin	Number
None given	27
Massachusetts	1,296
Connecticutt	318
Vermont	53
New Hampshire	29
Rhode Island	102
Ohio	1
New York	46
Florida	1
All American-born	1,873
Canada	452
Germany	98
Switzerland	3
England	78
Scotland	12
Ireland	375
Poland	6
Italy	1
All foreign-born	1,025
Total population	2,898

Source: Manuscript schedules, Eighth Census of the United States, 1860, Webster, Mass.

61. John F. Conlin, *Historical Sketches: A Retrospect of Fifty Years of St. Louis' Church, with Preliminary Chapters on the Early Days of Webster and Dudley* (Boston, 1901), pp. 31–32; Manuscript Schedules, Seventh Census of the United States, 1850; Eighth Census of the United States, 1860.

62. Conlin, *Historical Sketches*, pp. 31–32; Manuscript Schedules, Seventh Census of the United States, 1850; Webster, Mass., Vital Statistics, Births and Marriages, 1844–1860.

young workers. Men and women between the ages of sixteen and thirty-six entered the factories to tend the simple machines once operated almost exclusively by children and adolescents. The need for employment drove these people into the factories. Men preferred agricultural employment, and sometimes they were able to secure it in Webster. Irish laborers often moved back and forth between factory and fields, working a year or two in one place and a few months in another. William Scofield, for example, arrived in Webster around 1853 and worked as a laborer for several years before entering the mills, while Redmon Roach first worked as an operative, later was sent to the company farms, and still later returned to the factory.[63]

The Irish were not the only ethnic group to accept factory employment. Throughout the 1840s and 1850s they competed for work directly with French-Canadians, who began to arrive in the United States following the 1837–38 Montreal uprising. Population pressures on agricultural settlements in and around that city and a decline in the Quebec timber trade forced approximately 40,000 families to abandon their land and seek work in the United States. It has been estimated that in 1845 alone 22,000 immigrants arrived in the United States, and many of them found their way to the textile towns of New England.[64] This movement consisted generally of poor families, and not of individual itinerant workers. The proximity of Quebec and New England, the strong emotional and economic ties that bound kin together, and a pattern of family migration that was a characteristic of French-Canadian society encouraged this form of international movement. Arriving in the

63. Manuscript Schedules, Seventh Census of the United States, 1850; Eighth Census of the United States, 1860; Webster, Mass., Vital Statistics, Births and Marriages, 1844–1860.

64. The ratio of women to men was large enough to suggest that they immigrated in family units. Thirty-six percent of the migrants were under the age of sixteen, as well. See Frances Morehouse, "Canadian Migration in the Forties," *Canadian Historical Review* 9 (December 1928):326–328; Cole Harris, "Of Poverty and Helplessness in Petite-Nation," *Canadian Historical Review* 52 (March 1971):23–24; Jacques Henripin, "Population and Ecology: From Acceptance of Nature to Control: The Demography of the French Canadians since the Seventeenth Century," in *French Canadian Society*, ed. Marcel Rioux and Yves Martin, 2 vols. (1964), 1:204–210; and *Fourteenth Report of the Legislature of Massachusetts Relating to the Registry and Returns of Births, Marriages, and Deaths in the Commonwealth for the Years Ending December 31, 1855* (Boston, 1857), p. 166.

United States, French-Canadians drifted from one factory town to another before they secured steady employment.[65] Many of them settled in Webster.

Migratory families were large, averaging six or seven people, but couples with six, eight, and as many as twelve or thirteen children were not uncommon. Typical of Webster's new arrivals was Francis Laboisiere. "The bearer Francis Laboisiere not being able to talk English desires to take this method of asking you if you can find employment for him and for his family," wrote C. Macready of Fenner's factory to Samuel Slater and Sons in 1849. "He has 3 daughters one of 18 another of 15 and the other 14. He has also a son at 20."[66] If Slater required more workers, Laboisiere could find them; of course, Laboisiere and his wife could work in the mills as well. Another of Webster's new factory families was Augustus and Mary Tenor and their six children, who ranged in age from two to twenty. In 1858 they crossed over into New England and moved from one factory to another before they settled in Webster.[67]

In 1850 the French-Canadian and Irish populations were equal in size, and within the next decade French-Canadians became the largest ethnic group in Webster, comprising 16 percent of the total population and 24 percent of the entire labor force. Although movement into Webster was constant from the late 1840s onward, the greatest influx occurred between 1855 and 1860, when the number of new arrivals more than doubled. In the three quinquennial periods between 1845 and 1860 there were respectively 189, 196, and 452 French-Canadians living in Webster.[68]

When immigrants arrived in large numbers to fill positions in the factories, they moved into tenements newly constructed by the Slater family. Near the factories the company built forty-five new

65. Manuscript Schedules, Eighth Census of the United States, 1860.
66. Slater MSS, Samuel Slater and Sons, vol. 190, C. Macready to Storrs, February 26, 1849.
67. Manuscript Schedules, Eighth Census of the United States, 1860. See also *Thirteenth Annual Report of the Massachusetts Bureau of Statistics of Labor*, "The Canadian French in New England" (Boston, 1882), pp. 18–19; Vera Shlakman, "Economic History of a Factory Town: A Study of Chicopee, Massachusetts," *Smith College Studies in History* 20 (October 1934–July 1935):148.
68. *Abstract of the Census of the Commonwealth of Massachusetts with Reference to Facts Existing on the 1st Day of June, 1855*; see also Manuscript Schedules, Eighth Census of the United States, 1860.

block-style tenements, each of which accommodated from four to ten households. Built side by side along the roadway, these tenements preserved little of the appearance or the rural character of the dwellings the company had constructed in the 1820s and 1830s. These small, two-story wooden structures averaged about 28' by 39' for a three- or four-room tenement. They had fewer rooms and windows and less floor and storage space than earlier dwellings, and they provided almost no land for individual household gardens. Along Main and Pearl streets, few trees or open spaces interrupted the long rows of tenements. Most of these new units were built in clusters either next to or directly across from the factory. Physically these tenements and the factory now formed a unit distinct from the central village.[69] The immigrant community was isolated from the native-born commercial and residential sectors of town.

Changes were not confined to town planning and village architecture. The factory system made new demands on immigrant laborers. With their entry into the factory, the traditional division of labor based on gender and marital status, which had begun to break down under the Irish, disappeared altogether. Men marched into the factories. No longer could householders demand that Slater provide them with agricultural labor or with another job outside the factory and beyond the discipline of the factory agent. They needed work. Married women, too, entered the factory. French-Canadian women at every stage of the life cycle could be found in the mills. The Slater payroll records include the names of Charlotte Minor, age fifty-five and the mother of six; Merric Minor, age thirty-one and mother of two; and Philemon Dupra, age twenty-two and mother of two children under the age of five.[70] In open violation of truancy laws and factory acts, young children again entered the mills. Economic necessity forced parents to take factory work and even to deceive local factory agents about the ages

69. Beers, *Atlas of Worcester County*, pp. 93–94; Tolles, "Textile Mill Architecture," pp. 246–247; Fowler, "Rhode Island Mill Towns," p. 20; and Slater MSS, General Box 1, Samuel Slater and Sons to D. W. Jones, September 26, 1836.

70. *Abstract of the Census of the Commonwealth of Massachusetts with Reference to Facts Existing on the 1st Day of June, 1855*; Manuscript Schedules, Eighth Census of the United States, 1860; and Slater MSS, Union Mills, vol. 95, Families and Where Living and Where Employed, 1866.

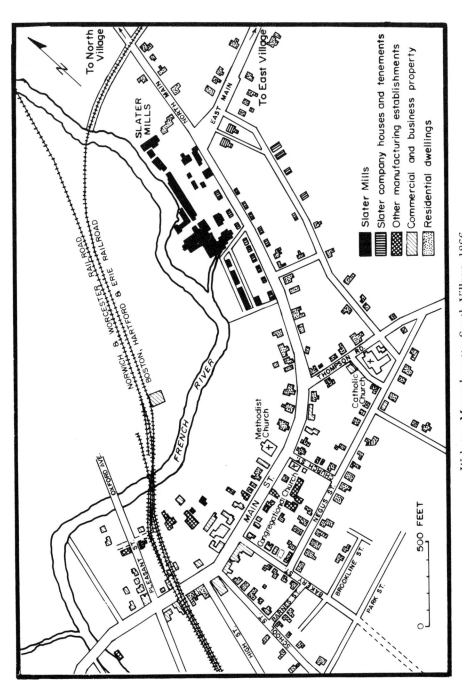

Webster, Massachusetts: South Village, 1866

of their children so that they, too, might enter the factories and contribute to the family income.[71] Altogether thirty-nine children twelve and under worked in the mills in 1860. Elia, Mary, Joseph, and Cylia Beshaw, who ranged in age from eight to thirteen, worked there; Joseph Minor, Merric's eleven-year-old son, worked beside them. Alphonzo Pockner and Michael Toony, age ten and twelve, joined them on their daily walk to the mills.[72] Laws designed to limit child labor proved insufficient when parents and factory masters conspired to circumvent them. "The most probable justification offered, is the oft repeated story of the poverty of the child's parents, and the absolute necessity of its earnings to protect against positive want," wrote H. K. Oliver in a report on the wages and working conditions of Massachusetts operatives. He blamed both the "personal self-interest of an unscrupulous employer, and . . . the poverty of the employed" for this situation.[73]

French-Canadian families became a cornerstone of Horatio Slater's new labor force. They comprised about 40 percent of the textile labor force. Those who entered the factories as operatives remained there through most of their working lives. In the wider community the situation was the same. Fully 96 percent of the French-Canadians in Webster performed unskilled work.[74] For these people, occupational mobility was almost nil: they entered at the bottom of the economic ladder and remained there.

The workers who entered the factories in the 1850s had little in common with those of an earlier generation. The customary barriers to employment at Samuel Slater and Sons, such as gender, age, marital status, and kinship ties, had disappeared. Men, women, and children, married or single, native-born or foreign-born, itinerant hands or family help, all accepted jobs operating the simple textile equipment. Within the factory, conditions also had changed: Horatio Nelson Slater and his assistants controlled the factory floor. Laborers worked within perimeters established by the managers. The hiring policies, the allocation of jobs, the discipline of workers

71. Ware, *Early New England Cotton Manufacture*, pp. 234–235; "Canadian French in New England," pp. 3–4.
72. Manuscript Schedules, Eighth Census of the United States, 1860; and Slater MSS, Union Mills, vol. 95, Families and Where Living and Where Employed, 1866.
73. *Report of the Hon. H. K. Oliver*, p. 20.
74. Manuscript Schedules, Eighth Census of the United States, 1860.

were management's prerogatives. With the breakdown of the earlier labor system in the 1840s and with the arrival of immigrant workers the following decade, a new relationship had to be established between labor and management. While French-Canadians and Irish would later refashion the workplace to meet many of their personal requirements, changes remained to be negotiated. The factory system established by Samuel Slater in 1790 had been transformed by his sons. Economic conditions and contemporary social values now dictated business policies.

Conclusion

While Slater's industrial communities often appeared to be peaceful, ordered societies, during periods of change the strains and contradictions that had been suppressed or dismissed by both Slater and male householders rose to the surface. There is little doubt that tension sometimes ran high, but it is more difficult to determine the locus of struggle. For some historians a class analysis explains all, but to push class struggle and the formation of class consciousness to the center of an analysis of the Slater system is to underestimate the cultural forces at work and to minimize the complexity of social relations. To a degree the Slater system represents a transition period preceding the development of class relations and class consciousness. To the people involved, both labor and management, class was not a salient world view, not a category in which people easily understood themselves and their environment. Power was exercised along gender and religious lines. Men and certain religious groups held power alongside Samuel Slater and his sons, and this common culture bound them together and overrode class interests. For laborers in these industrial towns, the hot friction points in their lives occurred in the relationships between men and women, parents and children, Protestants and Catholics.

STRIKES

There is clear evidence that the relationship between labor and management deteriorated substantially in the 1840s and 1850s and that some workers took collective action against manufacturers. But these confrontations did not necessarily constitute the emergence of new solidarities and new allegiances.

In southern New England, strikes occurred in family-style mills first in Fall River and then in Three Rivers in 1851, in Adams and Salisbury in 1852, and in Blackstone in 1853. In the Blackstone strike, workers demanded a 10 percent increase in wages, and when the Blackstone Manufacturing Company refused, some 700 operatives walked out. The firm closed down and remained closed for six months.[1] The strike in Salisbury centered on different issues. For more than twenty-five years employees of the Salisbury Manufacturing Company had enjoyed a fifteen-minute break in the morning and another in the afternoon. But a new company agent hired in 1852 believed that employees abused the privilege by absenting themselves from the mill for long periods and thus disrupting operations. In June he abolished the breaks and dismissed about a hundred operatives who refused to comply with his directive. This action aroused the anger both of the remaining operatives and of local residents, who joined workers in protest against the firm. Their forces were augmented by the operatives of the nearby Amesbury Flannel Mill, and a strike ensued.[2]

The actions taken by the manufacturers affected the entire community. In a series of resolutions dated June 5, 1852, the residents of Salisbury expressed their anger and dismay at the actions taken by the companies. They couched their grievances in traditional terms that could have been understood easily by Slater's native-born workers. The Salisbury operatives were appalled by "the withdrawal of privileges enjoyed for a quarter of a century, and which we looked upon as the settled and common law of our manufacturing establishments." They believed that

the operatives in this village have duly appreciated this privilege, and have been faithful and punctual in the discharge of their duties, and it is not too much to say, that as a class, in point of morality, industry and efficiency, they would bear a favorable comparison with any es-

1. *Report on Statistics of Labor* (1880), pt. 1, "Strikes in Massachusetts," p. 16; see also pp. 9–13.
2. Ibid. See also Merrill, *History of Amesbury*, p. 366.

tablishment in the Union. They are mostly permanent residents, under the wholesome influence of home; they have something at stake in the common prosperity; are good citizens in all the relations of life: and nowhere has law and order been better observed, and property more secure, than in this village.

Workers could not understand why management altered and abolished traditional prerogatives. The customary regulations, "so beneficial and honorable to both employers and employee, should not be lightly changed. It cannot be good policy to lose all the moral and social advantages which these establishments unquestionably have enjoyed over many others in New England. It cannot be well to array the interests of our village and those of the Corporation, against each other, or to fling new elements of discord and hatred into social and political life."[3] After a protracted struggle, management prevailed; the mills started up again, but this time with immigrant labor.[4]

Protests continued throughout the industry. In 1857 and 1858 the textile industry experienced a slump, and many factories closed down. Those that remained open often operated at half their normal capacity, and many tried to reduce costs by cutting wages. Workers in Salem, Newburyport, West Springfield, Adams, Blackstone, Uxbridge, and Webster protested. The Webster strike began in June 1858. Royal Storrs, agent at Webster, reported to the Providence office that production schedules were down because of "some disaffection . . . yesterday by the weavers. Some 25 to 30 having turned out for increase of wages."[5] The strike lasted a few days, and on July 3 Storrs reported: "We are pleased to observe that the weavers as Exp. [expected] have reconsidered their position and

3. Thomas F. Currier, "Whittier and the Amesbury-Salisbury Strike," *New England Quarterly* 8 (March–December 1935): 106–108.
4. Paul Faler and Melvyn Dubofsky have noted that some nineteenth-century strikes resembled public demonstrations. Everyone in Salisbury took sides in the confrontation between labor and management, and local residents raised approximately $2,000 for the support of strikers. See Faler, "Workingmen, Mechanics, and Social Change," p. 463; Melvyn Dubofsky, "Origins of Working-Class Radicalism, 1890–1905," *Labor History* 7 (Spring 1966): 131–154. See also "Strikes in Massachusetts," pp. 9–13, and Merrill, *History of Amesbury*, p. 366.
5. Slater MSS, Union Mills, vol. 119, R. Storrs to Samuel Slater and Sons, June 30, 1858.

are now . . . at work."[6] They returned to their looms without an increase in pay. The *Webster Journal*, which began publication that year, supported management and tried to discourage further protests. The editor advised laborers that "those who can get work at moderate pay, may consider themselves well off, and those who have money laid by to carry them through the winter may be deemed comparatively rich."[7] People were reminded that if conditions in the industry deteriorated further, they, too, might have to join the unemployed men and women who were in search of work.[8]

While only a small number of Webster's laborers participated in the strike (approximately 6 percent of the textile labor force), and while the strike may have been part of broader labor unrest in the region, the turnout was important because it served to crystallize the new relationship that had emerged between labor and management. The force of tradition had helped to regulate the relationship between the laborer and the factory master, but now nothing intervened to protect one from the other. Economic activity had been isolated from its broader social context; labor had become a commodity and was free to compete in the growing market economy.[9]

It is tempting to point to the Webster walkout, to the other strikes that occurred that decade, and to a turnout in Pawtucket, Rhode Island, in 1824 as testimony to class struggle, class formation, and signs of emerging class consciousness. A conclusion could be advanced that class struggle was intrinsic to industrialization, as some have suggested, or that "domination always incites resistance," as others have argued.[10] Indeed, a rising sense of worker consciousness has been a dominant theme of recent literature on the process of early industrialization in New England. The actions of shoemak-

6. Ibid., Union Mills, vol. 113, Storrs to Samuel Slater and Sons, July 2, 1858.
7. *Webster Journal*, November 20, 1858.
8. The problems of the unemployed were known in the community. Letters of inquiry such as the following one from Harrison Bacon arrived frequently at the factory: "I am out of work an did not know what to do, I thot I would write to you to see if you had any work for me. the mills are all stopped here. I will come an do any thing you want work in the mill or drive team an my wife may tend warper or spool please give us work" (Slater MSS, Union Mills, vol. 121, Harrison Bacon to Storrs, October 3, 1857).
9. Polanyi, *Great Transformation*, pp. 76, 78–80.
10. Prude, *Coming of Industrial Order*, p. xii; Russell Jacoby, *Dialectic of Defeat: Contours of Western Marxism* (New York, 1982), p. 19.

ers in Lynn, operatives in Lowell, and weavers in Pawtucket have been interpreted within this broad context. On Pawtucket's experiment with textile production, Gary Kulik writes: "Class, and class consciousness, emerged in Pawtucket as the textile industry developed. A new and distinct class, consisting of textile mill owners or of wealthy merchants with textile investments, formed itself prior to 1820."[11] Undeniably questions concerning class consciousness are important. In some situations this is the most appropriate analytical framework for understanding the transition that took place in the lives and work of early New Englanders during the Industrial Revolution. In other cases this analysis fits less well.

In Slater's industrial communities, class struggle was often a minor part of everyday experience, blunted by cultural factors. This is not to say that disputes between labor and management were not a significant part of the early industrialization process. The relationship between labor and management was not one between equals, and basic contradictions between the interests of labor and those of management did exist. The power of workers was tenuous at best, and this became evident during periods of economic stress and change. But the existence of objective circumstances was not enough to forge a new identity. History and experience have shown us that even when workers' situations objectively call for revolt, they often do not take collective action. Class consciousness has never come automatically to workers. In Webster, Wilkinsonville, Slatersville, and other industrial communities, cultural factors served to mask the power relationship. The force of tradition, the identification with Samuel Slater and Sons, the tenacious loyalty of laborers to traditional values and institutions prevented most workers from taking a pragmatic view of the factory system and the Slater family. The Webster turnout of 1858 indicated a level of discontent and frustration among workers, but it was not the harbinger of a new identity based on class. Most workers never challenged the new economic order directly. Flight was the overwhelming response of workers to their changed economic and social positions. While flight can be considered a bold form of protest, it represents an individual response to an altered situation. Flight may help a

11. Kulik, "Pawtucket Village and the Strike of 1824," p. 8.

minority of workers, but in the end it does not change the existing order.[12]

TENSION IN SLATER COMMUNITIES

Much of the tension observed in Slater's industrial communities can be located in traditional relationships. Industrialization exacerbated anxieties long a part of the traditional social order. The culture people share, their personal habits, and their customary beliefs change slowly; traditional beliefs shaped the reaction of early industrial workers to change for a long time. The implication of these early struggles for labor, management, and the emerging industrial order will become clear only through further research and analysis. The following examples serve only to suggest that traditional relationships formed a significant basis for tension in this society, that they cut across class lines, fragmented the lower orders, and inhibited the formation of a new identity.

An argument can be advanced that gender and generational tension divided labor. Under the Slater system, industrialization did not destroy the family. The employment of kinship units served initially to strengthen the family and to buttress the authority of the paternal head of household. Slater succeeded in assembling a labor force in part because he allowed a generation of male householders to reinforce patriarchal power within their families. Yet the adaptation of the family unit in this particular form was not accomplished without increasing internal familial tension.

The company records and other literature on industrial workers contain hints that gender conflict was in the making, that the factory system served as an arena in which men and women tried to redefine notions of authority, independence, and equality between the sexes.

One issue that frequently emerged involved the allocation of jobs within the mill. For many women, the new factory system held out the promise of economic and personal independence. The case of

12. Moore, *Injustice*, p. 79.

Rebecca Cole, one of the first industrial workers employed by Slater in Pawtucket, is illustrative. In the 1790s she secured a well-paid, respected job in the factory, earned enough to support herself and help support her children, and saved sufficient funds to ensure her future welfare. Her daughter Hannah became a mule spinner—a highly paid, skilled, male-dominated occupation. It appears, however, that Hannah Cole was the only woman to hold such a job, and that few women achieved the economic independence attained by her mother.[13] As the factory system took root, as Slater turned to male householders for his supply of hands, the advancement of women in the work force was checked. In the division of labor that followed, only men occupied skilled and supervisory positions, while women were confined to low-paying, low-status jobs primarily in the carding and spinning rooms. The familial authority structure that placed women in a subordinate, dependent position was transferred from home to factory.

Throughout the early stages of industrialization, laborers remained sensitive to issues associated with female advancement. An indication of this concern comes from a memorandum sent by John Slater to his Wilkinsonville agent, Alexander Hodges. In 1836 Hodges resigned as mill agent, and when discussing his replacement, John Slater asked him to reassure "the help that wish to know who they are going to work under that it will undoubtedly be a *man* therefore they need not have any apprehensions on that score."[14]

This communication raises some provocative questions. Did laborers, especially unskilled men, fear a loss of their own influence, position, and authority within the factory, the family, and the wider society if women advanced in the factory system? How was fear translated into practice? How did traditional concepts of patriarchy effect the attitudes of men toward women's wage rates and promotion and the abolition of the family wage system? Manufacturers certainly believed there was a breach between men and women which could be exploited. By paying women low wages and by ex-

13. For a discussion of Rebecca Cole, see chap. 3; see also Kulik, "Beginnings of the Industrial Revolution in America," pp. 204–208.
14. Slater MSS, Sutton Manufacturing Company, vol. 47, John Slater to Alexander Hodges, February 19, 1836. This note suggests that women held supervisory positions in some factories and that they may have been allowed to learn a skill.

acerbating tensions between men and women, they could keep costs down.

The introduction of power looms and the employment of young women and girls to operate the new equipment spotlighted some of the gender-specific strains in the new factory system. In part the Pawtucket turnout of 1824 revolved around gender-related issues. At the center of the strike stood female weavers. In the 1820s many Pawtucket manufacturers (though not Almy, Brown, and Slater) installed power looms and hired young women to operate the new equipment. In 1824 they cut the wages of these workers and extended the workday another hour. This directive effected all laborers. The women protested. To defend their position and to gain support from laboring men and others in the community, manufacturers tried to point up the socially undesirable effects of higher wages for women. Manufacturers claimed that female weavers received, "on an average, about two dollars per week, over and above their board; this was generally considered to be extravagant wages, for young women, more particularly, at this time of general depression much more than they could obtain in any other employment, and out of proportion to the wages of other help." [15] This "other help" presumably meant adult male labor. High wages threatened to close the economic gap between unskilled men and adolescent girls, to place men and women on a more equal economic footing, and to increase women's leverage within the family. The economic independence promised women by their new status in the factory threatened to undermine male authority.

The support the women received from unskilled male workers is open to question. In the end they did not secure a return to their former wage scale; they suffered a 50-cent cut, the amount decided upon by manufacturers. [16] While the refusal of men to support the

15. Kulik, "Pawtucket Village and the Strike of 1824," pp. 5–22; See also *Manufacturers' and Farmers' Journal and Providence and Pawtucket Advertiser,* June 7, 1824, quoted in Gilbane, "Social History of Samuel Slater's Pawtucket," p. 268.

16. Reports of the Pawtucket strike emphasize the role of female weavers and their determination to resist pay cuts; newspapers and other accounts provide few details about their male supporters. The *Manufacturers' and Farmers' Journal and Providence and Pawtucket Advertiser,* May 31, 1824, and June 7, 1824, referred to those who supported the strike as a "tumultous crowd," "great numbers," and the "most unprincipled and disorderly part of the village" (quoted in Gilbane, "A Social History of Samuel Slater's Pawtucket," p. 268).

[257]

women's position would eventually result in the degradation of both, gender-specific tensions seem to have precluded full cooperation on the wider basis of class.

While in Pawtucket issues associated with gender exploded into public view, in most of the mill towns where family labor was employed such tension remained latent. Managerial regulations together with parental restraint and pressure prevented young women from realizing early economic independence. When independence became a possibility in the 1830s, tension surfaced and the resultant conflict exhibited both gender and generational dimensions. These problems could be observed in Slater's communities.

Slater delayed the introduction of the power loom for a decade and more after its usefulness had been proved. When he did introduce the machine into his factories, he tried to smooth the integration of large numbers of women into the labor force by enlisting the support of male householders. He recruited weavers first from among factory families, and when they failed to supply all of the operatives he required, he encouraged householders to recruit additional weavers, to provide the young women with housing, and to bring them under family government. Furthermore, Slater retained the traditional mode of wage payment. In Slater's industrial villages, the household was an income-pooling unit dependent in part on the contribution of children and charges. All wages earned by these operatives were funneled directly into the householder's account. These measures buttressed the position of the male householder in the factory and in the family and cut off a source of potential conflict between men and management. Discontent, however, lay just beneath the surface. In Slater's communities, gender conflict merged with generational tension to produce considerable distress for factory families. The scene of conflict shifted from factory to family. Under changing conditions, daughters and parents engaged in a debate to redefine their relationship.

As a group power-loom weavers were among the highest paid workers in the factory. Whether fourteen or twenty-four, all weavers received the same piece rates, and by the 1830s the wages they received brought them close to economic self-sufficiency. But without control over their earnings, independence was impossible. A decision made by the company around the mid-1830s to pay workers individually rather than on a family basis made any desire for independence a distinct possibility. Opportunity was there now for those who for one reason or another wanted to leave home. Jobs existed for women not only in family-style mills but also in Lowell factories.[17]

Because of the economic independence afforded young women by factory employment and the new mode of wage payment, the family became a locus of struggle. Daughters now possessed the leverage to exact concessions from parents and to request changes in family government. Parents and daughters attempted to renegotiate their respective duties, responsibilities, and obligations; children tried to set limits on authority and obedience. If an accommodation could not be made, they could leave and find lodging in local boardinghouses. In the 1840s and 1850s many chose that option. Of the eighty-four residents of the two Slater boardinghouses in 1850, forty-seven were women and girls. Approximately 42.5 percent of these female lodgers were young, from fifteen to twenty-one years of age.[18] It appears that technological change associated with the power loom was potentially subversive of patriarchy.

In the new industrial order the interests of men and women, parents and children, often clashed. Much of the discontent, however, remained hidden, and in Slater's communities it only rarely erupted into public view. This was not the case with respect to religion. The relationship between native-born Protestant residents and Catholic immigrants was tense. By the mid-1850s the tension could not be contained, and Protestants struck out against the newcomers.

One incident involved the construction of a Catholic church. On the edge of the residential district, on land donated by John Carney, an Irishman who had resided in Webster since 1839, St. Louis Roman Catholic Church was constructed between 1851 and 1853. Within a few years of its completion, it was the scene of the first major outbreak of violence in Webster when native-born residents marched on the church and tried to burn it down. Guards had to

17. See chap. 9.
18. Manuscript Schedules, Seventh Census of the United States, 1850; Anderson, *Family Structure in Nineteenth-Century Lancashire*, pp. 123–132.

be stationed around the building to protect it. "It is said that the arrival of the pipes for the organ caused the rumor to spread that guns and ammunition were being stored in the basement, and that the church had been transformed into a barracks."[19]

At the same time, residents took additional steps to regulate Catholic behavior. In March 1855, alarmed by the increase in the Catholic population and afraid that the values held by these new residents would threaten their concept of the good society, local townspeople passed a resolution that urged all parents to send children to the public schools. Aimed largely at the immigrant Catholic population, the resolution read, in part:

> Whereas the Legislature in establishing free Schools aimed to provide means whereby the young might become filled not only for duties of private life but also to become good Citizens, and useful members of society capable of . . . sustaining and perpetuating the institutions of a free Religion and a liberal government and as these objects cannot be secured without a competent knowledge of the principles on which they are founded—therefore—Resolved—That the School Committee only as shall in there [*sic*] judgement be qualified to give instruction in relation to the natural rights of man in the fundamental principles of a sound and rational morality and of a free government and to see that such principles are taught in the various schools.[20]

Discord proved almost inevitable as the Protestant-dominated school committee superimposed its form of "rational morality" on an increasing Catholic student population.[21]

Attacks on Catholics could be viewed from several different perspectives. Angry at worsening conditions in the factory and concerned about maintaining their position in the boot and shoe industry, some native-born workers may have treated Catholics as scapegoats and transferred to this group the anger and frustration they felt toward Samuel Slater and Sons and toward the economic dislocation and insecurity in their lives. Or this incident may have been part of the general anti-Catholic hysteria found throughout

19. "Two Hundred Years of Progress," p. 40.

20. Webster, Mass., Town Meetings, Minutes: March 5, 1855.

21. Bruce Laurie, *Working People of Philadelphia, 1800–1850* (Philadelphia, 1980), pp. 128–130; Michael Feldberg, *The Philadelphia Riots of 1844: A Study of Ethnic Conflict* (Westport, 1975), pp. 33, 78.

Massachusetts at that time. Nativism and bad times often went together.[22] These explanations are plausible but not altogether satisfactory. They dismiss or trivialize the religious beliefs held by workers. The identity of many Webster residents was bound up with their religious beliefs and their church. Deeply committed to Protestant values and to a lifestyle that revolved around Sunday observance, weekly prayer meetings, Sunday schools, and revivals, native-born residents may have viewed the arrival of Catholic immigrants as threats to their value system and to their concept of the good, moral society. Whether misplaced anger, nativism, or religious convictions caused these incidents, the result of such confrontations was devastating for workers. Such incidents drove a wedge between members of the laboring classes; passion engendered by the conflicts highlighted the differences between workers and made effective cooperation on the wider basis of class difficult.[23]

Conflict based on gender, family, and religion divided the residents of Slater's industrial communities. Issues associated with class struggle and class consciousness, however, remained buried. People do not change their conceptions of the world easily, and this was especially true of early industrial workers. New ideas were not readily embraced, especially when they contrasted sharply with traditional convictions. And under Slater's guidance people were encouraged to hold on to the past. Industrialization did not imply a break with custom, a destruction of culture. The early factory system formed by Slater and his workers did much to bridge past and present. Retention of traditional, preindustrial culture simultaneously facilitated labor's integration into the new industrial order and induced identification with management. The relationship between employer and employee cannot be explained simply in terms of fear, coercion, or exploitation. An analysis based on such factors minimizes both the complexity of social relationships and the role traditional culture played in forging the first American factory system.

22. Feldberg, *Philadelphia Riots of 1844*, pp. 41–77.
23. Clifford Geertz, *Interpretation of Cultures: Selected Essays* (New York, 1973), pp. 201–208.

Index

Library of Congress Cataloging in Publication Data

Tucker, Barbara M.
   Samuel Slater and the origins of the American textile industry, 1790–1860.

   Includes index.
   1. Slater, Samuel, 1768–1835.   2. Textile industry—United States—Biography.   3.Textile industry—United States—History. I. Title.
HD9860.S5T83   1984   677'.21'0924 [B]   84-45145
ISBN 0-8014-1594-2 (alk. paper)